International Conference on Defects in Insulating Materials (ICDIM 2016)

IOP Conference Series: Materials Science and Engineering
Volume 169

Lyon, France
10 – 15 July 2016

Editors:

C. Dujardin
E. Homeyer
J. Houel
G. Ledoux

ISBN: 978-1-5108-3673-0
ISSN: 1757-8981

Printed from e-media with permission by:

Curran Associates, Inc.
57 Morehouse Lane
Red Hook, NY 12571

Some format issues inherent in the e-media version may also appear in this print version.

Printed by Curran Associates, Inc. (2017)

For permission requests, please contact the Institute of Physics
at the address below.

Institute of Physics
Dirac House, Temple Back
Bristol BS1 6BE UK

Phone: 44 1 17 929 7481
Fax: 44 1 17 920 0979

techtracking@iop.org

Additional copies of this publication are available from:

Curran Associates, Inc.
57 Morehouse Lane
Red Hook, NY 12571 USA
Phone: 845-758-0400
Fax: 845-758-2633
Email: curran@proceedings.com
Web: www.proceedings.com

Table of contents

Volume 169

2016 International Conference on Defects in Insulating Materials (ICDIM 2016)

10–15 July 2016, Lyon, France

Accepted papers received: 19 January 2017
Published online: 16 February 2017

Preface

Papers

IOP Conf. Series: Materials Science and Engineering **169** (2017) 011001 doi:10.1088/1757-899X/169/1/011001

Preface

C.Dujardin*, E.Homeyer, J.Houel G.Ledoux

Univ Lyon, Université Claude Bernard Lyon 1, CNRS, Institut Lumière Matière, F-69622 Villeurbanne, France

*Email: christophe.dujardin@univ-lyon1.fr

The 2016 edition of the International Conference on Defects in Insulating Materials (ICDIM 2016) was held in Lyon, France on July 10-15, 2016. A scientific community always originates from a common historical background and our common background is the science of color centers, which has been nicely reviewed by Prof. Heiz Pick in 1980 [1]. Because a scientific community always needs events to gather scientists from all around the world, setting up a conference series is fruitful way to regularly stimulate joint research and share recent advances. In this spirit, the first international event related to color center science was held in Argonne-USA in 1956. In 1973, a similar conference series entitled Europhysics Conference: Lattice defects in Ionic Crystals emerged in Europe (Marseille). In 1977, in Gatlingburg, the Colour Centre conference series became the first official International Conference on Defects in Insulating Materials and in 1994, in Lyon, the European meeting on Lattice Defects in Ionic Crystal merged with ICDIM and became EuroDIM. Both events occur every 4 years, leading to an event on materials defect science every 2 years. Figure 1 depicts the historical revue of the evolution of the ICDIM/EuroDIM conference series. Note that during ICDIM 2016, Dr. Martin Nikl and Prof. Rob Jackson have been elected as the chairperson of respectively ICDIM and EuroDIM conference series.

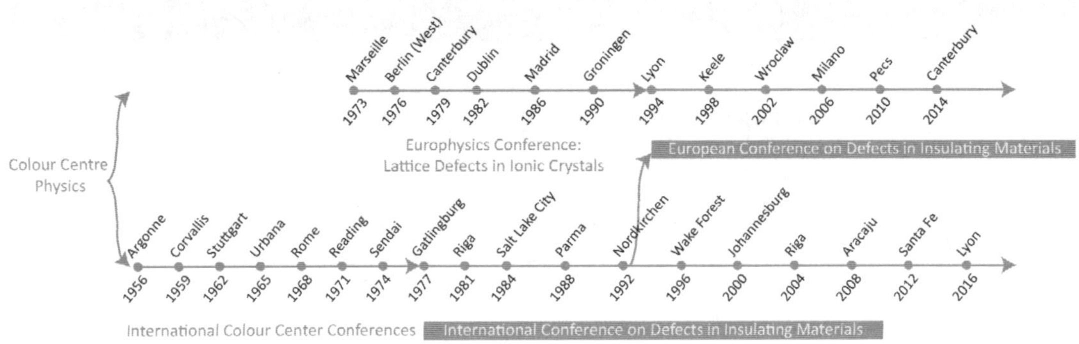

Figure 1: the historical evolution of the 60 years of conferences on defects in insulating crystals

This conference series is rather focused and aims to promote material sciences around defects, particularly in wide band gap crystals and glasses. But to adapt to the current trends, the content of the topics has evolved along the following items:
-Point and extended defects

-Phenomena at surfaces and interfaces
-Low dimensional and nano-morphological systems
-Defect modeling and computational methods
-Luminescence of excitons, impurities and defects
-Radiation induced defects and material damage
-Scintillation, energy transfer and storage, carrier trapping
-Laser actived centers and nonlinear optical phenomena
-Disordered systems
-Applied materials and interdisciplinary topics.

This issue of the Institute Of Physics conference series is dedicated to papers submitted during the ICDIM 2016 event in Lyon, France. 209 abstracts were selected for oral or poster presentation, always in plenary session. We have also proposed 7 invited keynote speakers who presented didactical talks introducing their respective fields to young researchers. Most of the talks have been video recorded and can be seen on the following YouTube channel [2].
The International Committee has selected Guangzhou (China) for ICDIM 2020, which will be chaired by Prof. Mingying Peng.

[1] H.Pick 1980, *J. de Physique, Colloque C6, Tome 41*, C6-1, Fifty years of colour centre physics
[2] https://www.youtube.com/playlist?list=PL_ZZHWrfptdQPjO8Q9H6rOBkuFS7pbTBV

Guest Editors
Prof. C.Dujardin, Conference Chair and Senior Editor
Dr. E.Homeyer, Associate Editor
Dr. J.Houel, Associate Editor
Dr. G.Ledoux, Associate Editor

Peer review statement

All papers published in this volume of *IOP Conference Series: Materials Science and Engineering* have been peer reviewed through processes administered by the proceedings Editors. Reviews were conducted by expert referees to the professional and scientific standards expected of a proceedings journal published by IOP Publishing.

IOP Conf. Series: Materials Science and Engineering **169** (2017) 012001 doi:10.1088/1757-899X/169/1/012001

Beyond lithium-ion batteries: A computational study on Na-S and Na-O batteries

[1,2] Masedi M C, [1]Ngoepe P E and [2] Sithole H M

[1]Materials Modelling Centre, University of Limpopo, School of Physical and Minerals Sciences, Private Bag x1106, Sovenga, 0727,
[2]CSIR, Meraka Institute, Meiring Naude, Brummeria, P. O. Box 395, Pretoria 0001, South Africa

cmasedi@csir.co.za

Abstract. The first principle pseudopotential calculations based on the Perdew-Burke-Ernzerhof (PBE) form of generalized gradient approximation (GGA) within density functional theory (DFT) has been utilized to investigate the stabilities of insoluble discharge products of oxygen and sulphur in the Na-O and Na-S batteries. Their structural, mechanical and electronic properties were determined. The lattice parameters were well reproduced and agree with the available experimental data. The heats of formation predict that all structures are generally stable and Na_2S has the lowest value. The elastic constants suggest that all the structures are mechanically stable which in good agreement with the calculated phonon dispersions.

1. Introduction.

Lithium-air batteries are potentially viable ultrahigh energy density chemical power sources, which could potentially offer specific energy up to 3000 Wh/kg being rechargeable [1]. Although their implementation holds the greatest promise in a number of applications ranging from portable electronics to electric vehicles (EV), there are also impressive challenges in developments of cathode materials and electrolyte system of these batteries. Li-air batteries has a number of difficult problems to overcome, the biggest shortfall exhibited with these systems is the formation of lithium dendrite which raises safety issues [2].

It has been suggested to replace the metallic lithium anode by sodium and operate the sodium-air cell, which could enable the development of a new generation of high specific energy rechargeable batteries. The theoretical specific energy of the sodium–air cell, assuming Na_2O as one of the discharge product and including the weight of oxygen, is 1690 Wh/kg, about four times that of state-of-the-art lithium-ion batteries. The surface tension of the liquid sodium anode is expected to prevent the formation of sodium dendrites on charge. Any sodium dendrites that might be formed would be absorbed into the liquid phase [3].

In the current work we present a comparative study on stability, structural and electronic properties of discharge products of sulphur and oxygen in Na-O and Na-S batteries.

IOP Conf. Series: Materials Science and Engineering **169** (2017) 012001 doi:10.1088/1757-899X/169/1/012001

2. Structures.

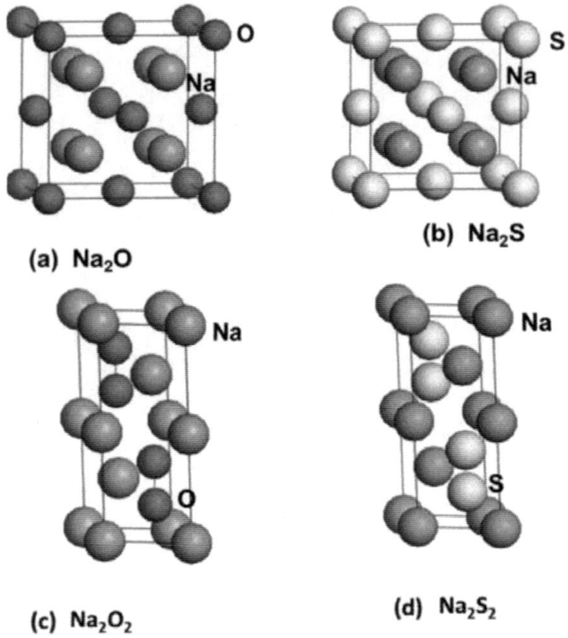

Figure 1: Crystal structures of Na_2S, Na_2O, Na_2O_2 and Na_2S_2, the red atom represent O, yellow atom is S and purple atom is Na.

Figure 1 shows the crystal structures of Na_2S, Na_2O, Na_2O_2 and Na_2S_2 to be studied in this work. Na_2S and Na_2O both have a cubic anti-fluorite structure with Fm-3m symmetry whereas Na_2O_2 and Na_2S_2 show a hexagonal structure with symmetry P_{63}/mmc. There is no much study performed on Na_2O_2 and Na_2S_2 both computationally and experimentally. Na_2O_2 and Na_2S_2 systems were generated from existing Li_2O_2 by replacing lithium with sodium and oxygen with sulphur using VASP total energy package [4]

3. Methodology.

The calculations were carried out using *ab initio* density functional theory (DFT) formalism as implemented in the VASP total energy package [4] with the projector augmented wave (PAW) [5]. An energy cutoff of 500 eV was used, as it was sufficient to converge the total energy of all the systems. For the exchange-correlation functional, the generalized gradient approximation of Perdew and Wang (GGA-PBE) [6] was chosen. The Brillouin zone integrations were performed for suitably large sets of k points according to Monkhorst and Pack [7]. The phonon dispersion spectra were evaluated using PHONON code [8] as implemented by Materials Design within their MedeA software, VASP code [4]. k-point mesh of 8x8x8 was used. Optimization of structural parameters (atomic positions and lattice parameters) was achieved by minimization of forces and stress tensors.

The heat of formation can be estimated by

$$\Delta H_f = E_c - \sum_i x_i E_i \qquad (2\text{-}21)$$

where E_c is the calculated total energy of the compound, E_i is the calculated total energy of element i in the compound.

4. Results and Discussions.

4.1. Structural Properties.

Table 1: The equilibrium lattice parameters and heats of formation (ΔH_f) of the Na_2O, Na_2S, Na_2O_2 and Na_2S_2 structures.

Structure	Lattice Parameters (Å)		ΔH_f (KJ/mol)		Volume
	VASP	Exp	VASP	Exp	(Å^3)
Na_2O	a=5.12	5.49 [9]	-318.25		134.29
Na_2S	a= 6.55	6.52 [10]	-298.13		281.50
Na_2O_2	a=5.26 c=4.45		-145.08		107.06
Na_2S_2	a=7.66 c=5.3		-148.11		274.26

There is good agreement between the experimental and calculated lattice constants, especially for Na_2S and Na_2O agrees within 7%. The observed values of heats of formation (ΔH) suggest that all structures are generally stable which is in good agreement with calculated phonon dispersions. Na_2O has the lowest heat of formation.

4.2. Phonon Dispersions Calculations.

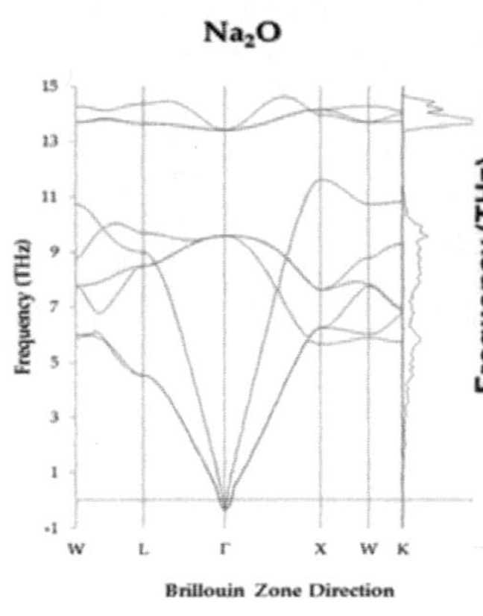

This work

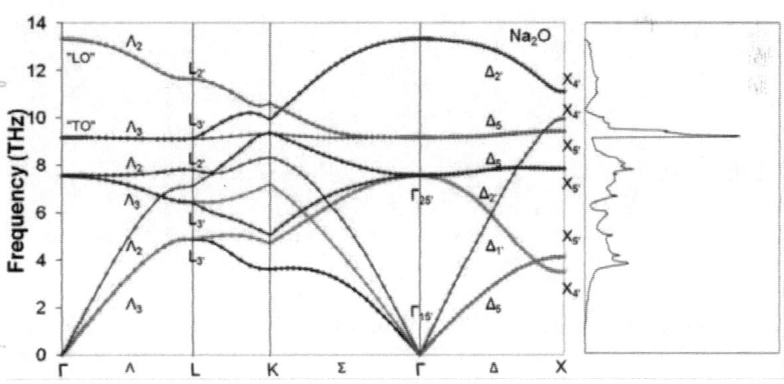

Literature [11]

IOP Conf. Series: Materials Science and Engineering **169** (2017) 012001 doi:10.1088/1757-899X/169/1/012001

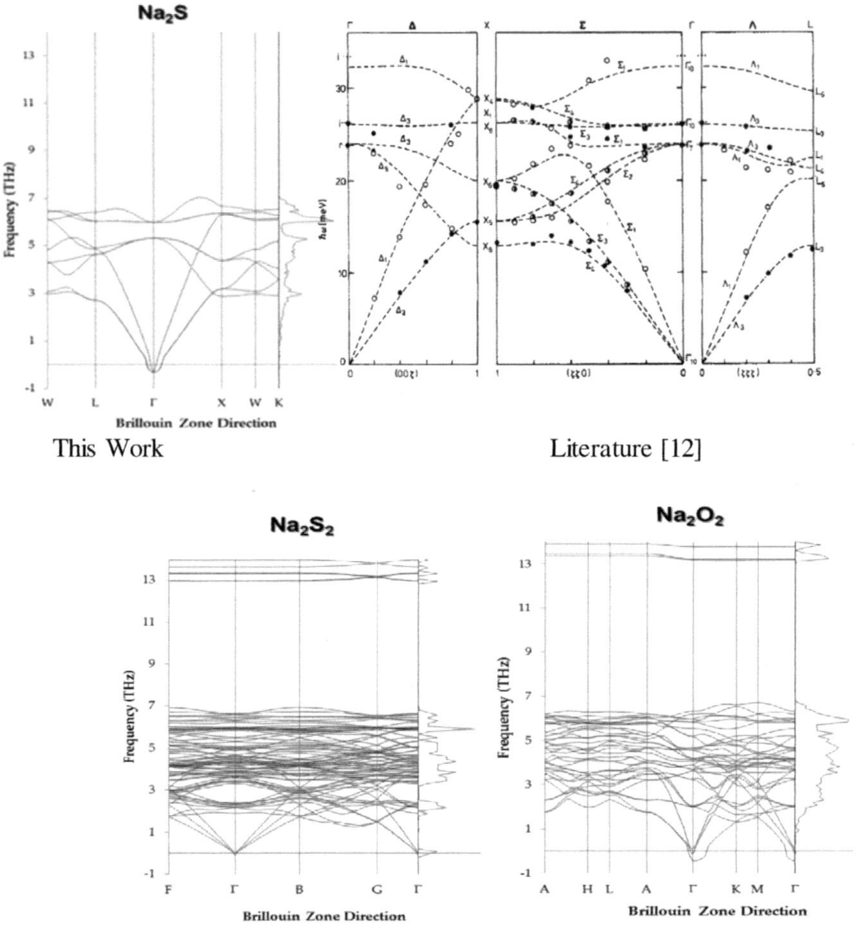

Figure 2: Phonon dispersion curves of Na_2S, Na_2O, Na_2O_2 and Na_2S_2 structures.

The phonon dispersion calculations show that all the structures are stable due absence or minimum soft modes observed in the along gamma point of the Brillouin zone and is in good agreement with the calculated elastic properties.

The phonon dispersion Na_2S and Na_2O compares well with some of work done with DFT and experiment, whereas for Na_2O_2 and Na_2S_2 there is no much work on phonon dispersions.

4.3. Elastic Properties.

Table 2: The elastic constants (GPa) for Na_2O, Na_2S, Na_2O_2 and Na_2S_2 structures.

	Na_2O		Na_2S		Na_2O_2	Na_2S_2
	VASP	Exp [13]	VASP	Exp [10]	VASP	VASP
C_{11}	127.59	114.00	51.61		220.40	59.35
C_{12}	75.87	34.71	16.25		147.98	16.78
C_{13}					101.21	12.14
C_{33}					263.25	59.88
C_{44}	26.67	27.40	14.56		79.57	9.99

The accurate calculation of elasticity is essential for gaining an insight into the mechanical stability and elastic properties of solids. The mechanical stability criteria of cubic systems as outlined elsewhere [14] [15] [16]are given as follows:

$$C_{44}>0, \quad C_{11} > C_{44}, \quad C_{11}+2C_{12}>0 \qquad (1)$$

where C_{11}, C_{12} and C_{44} are the only three independent elastic constants. The corresponding mechanical stability criterion for hexagonal crystal reads as

$$C_{11}>0, \quad C_{11}-C_{12}>0, C_{44}>0, \quad (C_{11}-C_{12})\,C_{33}-2C_{13}^{2}>0 \qquad (2)$$

Na_2O, Na_2S, Na_2O_2 and Na_2S_2, are positive, thus satisfying mechanical stability criteria set in equations (1) and (2).
Which means suggest that Na_2O, Na_2S, Na_2O_2 and Na_2S_2 are mechanically stable.

Conclusion.

Heats of formations suggest that all structures (Na_2O, Na_2S, Na_2O_2 and Na_2S_2) are stable because of negative values of heats of formation. The calculated values of lattice parameters and elastic constants are in reasonable agreement with the available experimental values for both Na_2O and Na_2S. The elastic constants suggest mechanical stability for cubic systems Na_2O and Na_2S and hexagonal systems Na_2O_2 and Na_2S_2 since their necessary mechanical stability conditions are satisfied. The phonon dispersions calculations show that all structures are stable which is further confirmed by positive elastic properties values and is in good agreement with some work done in literature.

Acknowledgements

This work was supported by the Council for the Scientific and Industrial Research (CSIR). The computations were performed at the Materials Modelling Centre (MMC), University of Limpopo and Center for High Performance Computing (CHPC), South Africa. We also acknowledge the support of the South Africa Research Chair Initiative of the Department of Science and Technology and the National Research Foundation

References

[1] Abrahamas K M and Jiang Z, *J. Electrochem. Soc.,* vol. 143, p. 1, 1996.

[2] Armstrong A R, Holzapfel M, Novák P, Johnson C S, Kang S H, Thackeray M M and Bruce P G, *J Am Chem Soc.,* vol. 128, p. 26, 2006.

[3] Peled E, *J. Electrochem. Soc.,* vol. 126, p. 2047, 1979.

[4] Blöchl P E, *Phys. Rev.,* vol. B 50, p. 17953, 1994.

IOP Conf. Series: Materials Science and Engineering **169** (2017) 012001 doi:10.1088/1757-899X/169/1/012001

[5] Wang Y and Perdew J P, *Phys. Rev.,* vol. B 45, p. 13244, 1992.

[6] Pack J D and Monkhorst H J, *Phys. Rev.,* vol. B 13, p. 5188, 1976.

[7] Parlinski K, Li Z Q and Kawazoe Y, *Phys. Rev. Lett.,* vol. 78, p. 4063, 1997.

[8] Kamran S, Chen K and Chen L, *Phys. Rev.,* vol. B 79, p. 024106, 2009.

[9] Zintl E, Harder A and Dauth B, *Z Elektrochem,* vol. 40, p. 588, 1934.

[10] H. Khachai, R. Khenata, A. Bouhemadou, H. Reshak Ali, A. Haddou, M. Rabah and B. Soudini, *Solid State Commun,* vol. 147, p. 178, 2008.

[11] M. Thompson,X. Shen and P.B. Allen, "Density functional calculation of electronic structure and phonon spectra of Na2O," *Phys. Rev. B ,* vol. 79, p. 113108, 2009.

[12] W. Buhrer and H. Bill, "Lattice dynamics of Na,S," *J. Phys. C: Solid St. Phys.,* vol. 13, p. 5495, 1980.

[13] Thomson M, Shen X and Allen B, *Physical Review,* vol. B79, p. 113108, 2009.

[14] Chauke H R, *PhD Thesis, University of Limpopo,* 2005.

[15] S. Kamran, K. Chen and L. Chen, *Phys. Rev. B,* vol. 79, p. 024106 , 2009.

[16] H. M. a. J. Pack, vol. B 13, 1976.

[17]]. J. P. P. a. Y. Wang, vol. B 45, 1992.

[18] B. Dunn, H. Kamath, J.-M. Tarascon, *Science 334,* pp. 928-935, 2011.

[19] P.G. Bruce, S.A. Frauberger, L.J. Hardwick, J.-M. Tarascon, *Nat. Mater. 11,* pp. 19-29, 2012.

[20] M. Armand, J.-M. Tarascon, *Nature 451,* pp. 652-657, 2008.

[21] J.-M. Tarascon and M. Armand, *Nature 414 ,* pp. 359-367, 2001.

2016 International Conference on Defects in Insulating Materials (ICDIM 2016) IOP Publishing
IOP Conf. Series: Materials Science and Engineering **169** (2017) 012002 doi:10.1088/1757-899X/169/1/012002

The importance of polarizability in the modeling of ionic diffusion in ceria

A K Lucid[1], **G W Watson**[1]

[1]School of Chemistry and CRANN, Trinity College Dublin, College Green, Dublin 2, Ireland

E-mail: lucida@tcd.ie, watsong@tcd.ie

Abstract. Classical molecular dynamics can be used to investigate ionic diffusion and its limitations in trivalently doped ceria and at the surfaces and interfaces of these materials. Here we compare the performance of two interatomic potentials derived for samarium doped ceria from the same set of ab-initio data, a dipole polarizable ion model (DIPPIM) and a rigid ion model (RIM). The DIPPIM allows for polarization effects resulting from induced dipoles whereas the RIM does not. In this study we aim to elucidate whether or not this system can be modelled successfully using a RIM or if a DIPPIM is necessary due to the large polarization effects caused by the presence of oxide (O^{2-}) ions.

1. Introduction

Oxide ions, which are the charge carriers in oxide ion conducting materials, are highly polarizable which is often ignored when modeling these materials.[1, 2, 3] Polarizability can be accounted for in interatomic potentials in a number of ways; shell models[4], embedded atom models[5, 6] and models where dipoles are solved self consistently as a response to changes in electric field[7] can all be utilized. These polarizable models can be computationally expensive and very time consuming, particularly those which calculate dipoles self consistently, and so it is worthwhile to determine whether or not the polarizability of the oxide ions must be taken into account.

Solid oxide fuel cells (SOFCs) are an example of a device in which oxide ion conduction is essential to the performance of the system. These devices offer an alternative for efficient, clean and renewable energy conversion in this environmentally conscious era.[8] These devices currently require high temperatures (~1273K) to operate; this is necessary for oxide ion diffusion in the electrolyte to occur. The most commonly used electrolyte materials in these devices is yttria-stabilized zirconia(YSZ). These high temperatures contribute to the cost, performance degradation and inefficiency of current SOFCs. Ceria (CeO_2) doped with trivalent cations has been suggested as a material which could replace YSZ as the solid electrolyte as it can achieve ionic conductivities comparable to YSZ but in the intermediate temperature (IT) range of 773-1073K. [9] Samarium (Sm) and gadolinium (Gd) are the dopants which have been seen to perform best with ceria as SOFC electrolytes.[10, 11] In order to optimize these materials for applications (i.e. to understand behaviors at surfaces and interfaces) we need to have a model which can accurately calculate the conductivities and diffusion mechanisms in these systems.

Here we present a comparison of a rigid ion model (RIM) and a dipole polarizable ion model (DIPPIM)[7] which have both been fitted to the same set of ab initio data[12]; the RIM is ~10 times less computationally demanding than the DIPPIM. However, the DIPPIM is orders

IOP Conf. Series: Materials Science and Engineering **169** (2017) 012002 doi:10.1088/1757-899X/169/1/012002

of magnitude less computationally expensive than ab initio MD; ab initio MD would not be tractable for the size of cells which we use to perform calculations. By fitting to ab initio potential energy surfaces, with accurate inclusion of polarizability, DIPPIM allows ab initio quality MD for a fraction of the cost. The testing was carried out on Sm-doped CeO_2 and pure CeO_2 by investigation of ionic conductivity, activation energy, structure, defect interactions and phonon modes.

2. The interatomic potentials

The RIM and DIPPIM interatomic potentials used here have been fitted to identical force, stress and dipole data which was generated using hybrid-DFT calculations.[12] They both include a pair potential, which is comprised of Coulombic interactions, a modified Buckingham type short-range repulsive term and a dispersion term, as described by Castiglione *et al.*[7] In the DIPPIM, polarization effects resulting from induced dipoles on ions in the system are taken into account. The part of the potential which is associated with the polarization includes only dipolar effects. The induced dipole on each ion for a given configuration is obtained by minimization, at each MD step, of the following potential with respect to all dipoles ($\boldsymbol{\mu}_i$):

$$U^{pol} = \sum_{j\neq i}^{N}[f_{ij}^{*}\mathbf{T}^{(1)}(\mathbf{r}_{ij}) \cdot (q_j\boldsymbol{\mu}_i - q_i\boldsymbol{\mu}_j) + \sum_{j\neq i}^{N}\boldsymbol{\mu}_i \cdot \mathbf{T}^{(2)}(\mathbf{r}_{ij}) \cdot \boldsymbol{\mu}_j] + \sum_{i=1}^{N}\frac{1}{2\alpha_i}|\boldsymbol{\mu}_i|^2 \tag{1}$$

$$f_{ij}^{*} = 1 - c_{ij}^{*}e^{-b_{ij}^{*}r_{ij}}\sum_{k=0}^{4}\frac{(b_{ij}^{*}r_{ij})^k}{k!} \tag{2}$$

Where q_i is the formal ionic charge on each ion, and the interaction tensors $\mathbf{T}^{(1)}(\mathbf{r}_{ij})$ and $\mathbf{T}^{(2)}(\mathbf{r}_{ij})$ are the charge-dipole and dipole-dipole interaction tensors. The first term in Equation 1 represents the charge-dipole interactions between atoms i and j. This interaction is damped by a Tang-Toennies function (f_{ij}^{*}, Equation 2), to account for short-range induction effects. The second term accounts for the dipole-dipole interactions and the final term is the potential energy of the polarization, where $\boldsymbol{\alpha}_i$ is the polarizability of the ion. This is the energy required to polarize the ion. It is the optimization of the dipole energy (Equation 1) which accounts for the computational expense of the DIPPIM.

3. Simulation details

Calculations were carried out using the finite temperature molecular dynamics program PIMAIM [13]. All simulations were carried out on 6×6×6 supercells of $Ce_{1-x}Sm_xO_{2-x/2}$ equating to ~2592 atoms depending the the dopant concentration. Calculations were carried out at five dopant concentrations (x = 0.05, 0.10, 0.15, 0.20, and 0.25) and at three temperatures, spanning the IT range (773K, 973K, and 1173K). To account for the statistical nature of MD, three randomized configurations of dopants and oxygen vacancies were considered for each dopant concentration. As a result of running calculations on three configurations values such as ionic conductivity and activation energy can be reported as average values.

Each system was first optimized, and temperature scaling was then carried out every 0.025ps for 10ps. The supercells were allowed 40ps to come to thermal equilibrium before data collection for analysis began. The data collection runs were 3ns for 773K and 1ns for 973K and 1173K. An isothermal-isobaric ensemble (*NPT*)[14] was used, with a timestep of 4fs. The short-range cutoff for the interatomic potentials and the cutoff for the Ewald sum was 11Å.

4. Results and discussion

4.1. Ionic conductivity and activation energy

The diffusion coefficient, D, was calculated from the mean squared displacement (MSD) of the trajectory of the oxide ions and was used to calculate the ionic conductivity from the Nernst-Einstein equation:

$$\sigma = \left(\frac{\rho q^2}{H_R k_B T}\right) D \tag{3}$$

where ρ is the number density of the charge carriers (in this case these are the oxide ions), q is the effective charge of the charge carriers, H_R is the Haven ratio (=1), k_B is Boltzmann's constant and T is the temperature of the system. The ionic conductivity, at 973K, for both the RIM and the DIPPIM is displayed in Figure 1(a), and it can be seen that there is at least an order of magnitude difference between the two models. The ionic conductivities for the RIM and DIPPIM are compared to experiment in Figure 1(b), with the experimental studies by Zha *et al.* and Jung *et al.* [11, 15] selected to show the variation in the literature, both in ionic conductivity and in the concentration at which the maximum in conductivity is seen. It can quite clearly be seen from Figure 1 that the DIPPIM is significantly better than the RIM at calculating ionic conductivities comparable to experiment which is vital to the performance of these materials as solid electrolytes.

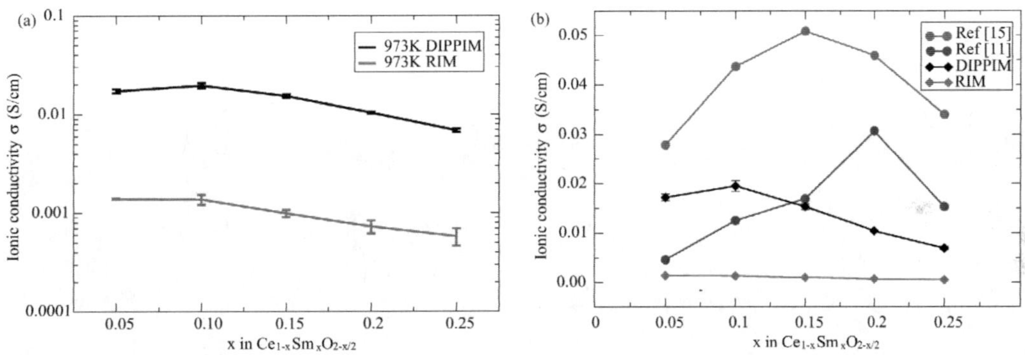

Figure 1. The ionic conductivity of $Ce_{1-x}Sm_xO_{2-x/2}$ at 973K for DIPPIM (black) and RIM (red) is shown in (a); it should be noted that this is a log scale. The ionic conductivity of the DIPPIM, RIM and two experimental studies[15, 11](green and blue) are shown at 973K in (b).

The activation energy for ionic conductivity can be calculated using:

$$\sigma T = \sigma_0 \exp\left(-\frac{E_a}{k_B T}\right) \tag{4}$$

where E_a is the activation energy, σ_0 is the attempt frequency of the hopping process, and all other terms are as previously defined. A plot of $\ln(\sigma T)$ versus $1/T$ was used to obtain the E_a of $Ce_{0.9}Sm_{0.1}O_{1.95}$ from 773K to 1173K (Table 1). Both the RIM and DIPPIM are in good agreement with experiment, suggesting that the activation energy of diffusion is being correctly described by both models. However, based on the results seen for ionic conductivity, it suggests that the RIM may not correctly model the attempt frequency of the hops.

Table 1. Activation energy for ionic conductivity of $Ce_{0.9}Sm_{0.1}O_{1.95}$ (E_a) for both the DIPPIM and RIM, compared to experiment.

E_a (eV)	Reference
0.610	DIPPIM
0.607	RIM
0.645	[11]
0.628	[16]
0.615	[17]

4.2. Structure

The partial radial distribution functions (RDFs) for Ce-O, Sm-O and O-O were all calculated; no major discrepancies were seen between the two models, with the RIM showing slightly sharper peaks which could be accounted for by the difference in dynamics between the RIM and DIPPIM. The dopant-vacancy (Sm-Vac) and vacancy-vacancy (Vac-Vac) RDFs were also calculated; these are shown for $Ce_{0.9}Sm_{0.1}O_{1.95}$ at 973K in Figure 2. The Ce-O and O-O RDFs are plotted with the Sm-Vac and Vac-Vac RDFs as these represent a random distribution of cations and vacancies when no ordering has occurred. On investigation of the Sm-Vac RDFs there is a very clear difference seen between the RIM and DIPPIM Sm-Vac distribution. For the RIM there is trapping of the vacancies in the nearest-neighbor site to the Sm ions. As the same is not seen for DIPPIM, it would appear that the RIM is severely overestimating the attraction between the dopant cations and vacancies which may be responsible for the severe drop in ionic conductivity that is seen when using the RIM. For the Vac-Vac RDF it can be seen that the DIPPIM predicts enhanced vacancy ordering in the $\langle 111 \rangle$ direction of the anion sub-lattice when compared with the RIM, although the long-range ordering is in agreement with the DIPPIM. This enhanced vacancy ordering in the $\langle 111 \rangle$ direction has been seen for doped fluorite materials previously and is believed to be the main reason for the drop in ionic conductivity of these materials past a certain vacancy concentration.[18, 19, 20]

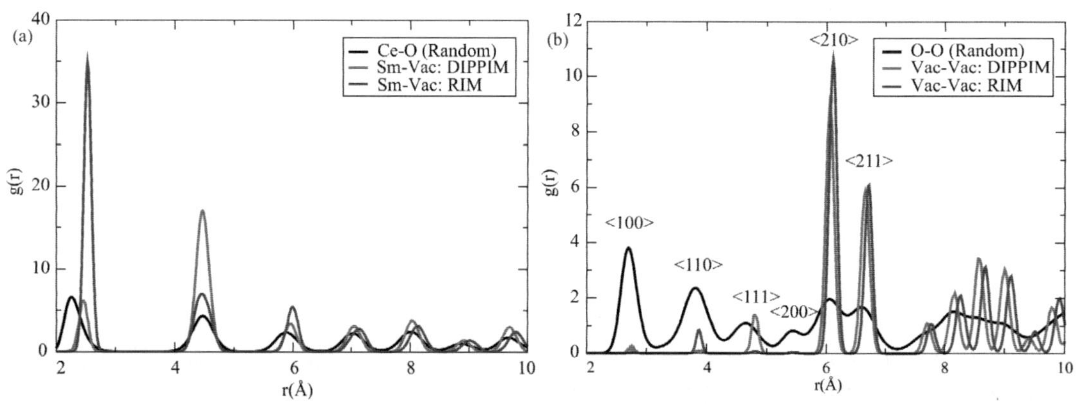

Figure 2. Sm-Vacancy (a) and Vacancy-Vacancy (b) partial radial distribution functions at 973K for $Ce_{0.9}Sm_{0.1}O_{1.95}$, for both the RIM and DIPPIM. The Sm-vacancy RDFs are plotted against the Ce-O RDF representing a randomized distribution of cations and vacancies. The vacancy-vacancy RDFs are plotted against the O-O RDF representing a randomized distribution of vacancies. The labels on peaks in (b) indicate directions in the simple cubic anion sublattice.

4.3. Phonon modes in pure CeO_2

Phonon modes in pure CeO_2 were also investigated as a test for the interatomic potentials. GULP[21] was used for the RIM and a combination of PIMAIM and phonopy[22] for DIPPIM, as DIPPIM is not implemented in GULP. The non-analytical correction (NAC) term, at the Γ-point[23, 24, 25], was included for the DIPPIM in order to determine the splitting between the transverse optical (TO) and longitudinal optical (LO) modes; this was not necessary for the RIM as GULP takes this correction into account. The Born effective charges, Z^*, were taken from Buckeridge *et al.*, with Z^* Ce $= 5.525$ and Z^* O $= 2.764$ (obtained using GGA+U)[26] and the high-frequency dielectric constant ($\epsilon^{\infty} = 5.31$) was taken from experiment[27]. This method was found by Buckeridge *et al* to obtain the values most comparable to experiment.

The calculated phonon mode frequencies are displayed in Table 2 for high symmetry points of the Brillouin zone, Γ, X and L; by definition the RIM will not correctly describe the LO-TO splitting as $\epsilon^{\infty} = 1$, so it is not surprising that the F_{1u} (LO) mode is overestimated by ~57%. It is of more interest to look at points away from Γ due to this failure to represent LO-TO splitting; the RIM tends to overestimate the phonon modes consistently and on average deviates more from experimental phonon modes when compared with the DIPPIM. This suggests that the RIM is failing to correctly reproduce the motions of the Ce and O ions in the system. This poor estimation of the frequencies by the RIM may lead to less attempted hopping in the system which could account for the poor calculation of ionic conductivity by this model.

Table 2. Calculated phonon mode frequencies of pure CeO_2 at high symmetry points of the Brillouin zone, Γ, X and L, for both the RIM and DIPPIM compared to experiment. Only modes which had available experimental comparisons are presented here. Units of frequency are cm^{-1}.

Mode	RIM	DIPPIM	Expt.
		Γ	
F_{1u} (TO)	291	274	218[28],274[29],275[28]
F_{2g}	524	490	460[29],465[30, 28]
F_{1u} (LO)	920	586	585[28],597[31]
		X	
E_u	187	123	138[29]
E_g	332	211	249[29]
E_u	635	503	464 [29]
		L	
E_u	142	107	138[29]
A_{1u}	256	217	227[29]

5. Conclusions

In conclusion, we have tested two interatomic potentials derived from the same set of ab initio data where one includes polarization effects and the other does not. We have seen that while both potentials reproduce experimental activation energies of oxide ion diffusion, the same can not be said for the ionic conductivity and dopant-vacancy interactions in Sm-doped CeO_2 or the phonon modes in pure CeO_2. Despite the fact that it is ~10 times more computationally expensive it is necessary to employ the DIPPIM in order to accurately model these materials for SOFC electrolyte applications, due to the large polarizability of the oxide ions.

Acknowledgments

This research was supported by Science Foundation Ireland (SFI) through the Investigators Programme (Grant No. 12/IA/1414). All calculations were performed using the Kelvin (funded through grants from the Higher Education Authority, through its PRTLI program), Lonsdale (funded through a grant from SFI - 06/IN.1/I92/EC07) and Pople (funded by SFI - 12/IA/1414) supercomputers maintained by the Trinity Centre for High Performance Computing (TCHPC), and the Fionn supercomputer, maintained by ICHEC (tcche051b).

References

[1] Williams N R, Molinari M, Parker S C and Storr M T 2015 *J. Nucl. Mater.* **458** 45–55
[2] Yoshiya M and Oyama T 2011 *J. Mater. Sci.* **46** 4176–4190
[3] González-Romero R L, Meléndez J J, Gómez-Garcia D, Cumbrera F L and Dominguez-Rodriguez A 2012 *Solid State Ionics* **219** 1–10
[4] Lewis G V and Catlow C R a 1985 *J. Phys. C Solid State Phys.* **18** 1149–1161

[5] Daw M S and Baskes M I 1984 *Phys. Rev. B* **29** 6443–6453
[6] Rushton M J D and Chroneos A 2014 *Sci. Rep.* **4** 6068
[7] Castiglione M J, Wilson M and Madden P a 1999 *J. Phys. Condens. Matter* **11** 9009–9024
[8] Stambouli A and Traversa E 2002 *Renewanable Sustainable Energy Rev.* **6** 433–455
[9] Kilner J A and Brook R J 1982 *Solid State Ionics* **6** 237–252
[10] Inaba H 1996 *Solid State Ionics* **83** 1–16
[11] Zha S, Xia C and Meng G 2003 *J. Power Sources* **115** 44–48
[12] Burbano M, Nadin S, Marrocchelli D, Salanne M and Watson G W 2014 *Phys. Chem. Chem. Phys.* **16** 8320–8331
[13] Madden P A and Wilson M 1996 *Chem. Soc. Rev.* **25** 339–350
[14] Martyna G J, Tobias D J and Klein M L 1994 *J. Chem. Phys.* **101** 4177–4189
[15] Jung G B, Huang T J and Chang C L 2002 *J. Solid State Electrochem.* **6** 225–230
[16] Omar S, Wachsman E D, Jones J L and Nino J C 2009 *J. Am. Ceram. Soc.* **92** 2674–2681
[17] Kasse R M and Nino J C 2013 *J. Alloys Compd.* **575** 399–402
[18] Marrocchelli D, Madden P A, Norberg S T and Hull S 2011 *Chem. Mater.* **23** 1365–1373
[19] Burbano M, Norberg S T, Hull S, Eriksson S G, Marrocchelli D, Madden P A and Watson G W 2012 *Chem. Mater.* **24** 222–229
[20] Hull S, Norberg S, Ahmed I, Eriksson S, Marrocchelli D and Madden P 2009 *J. Solid State Chem.* **182** 2815–2821
[21] Gale J D and Rohl a L 2003 *Mol. Simul.* **29** 291–341
[22] Togo A and Tanaka I 2015 *Scr. Mater.* **108** 1–5
[23] Pick R, Cohen M and Martin R 1970 *Phys. Rev. B* **1** 910–920
[24] Giannozzi P, De Gironcoli S, Pavone P and Baroni S 1991 *Phys. Rev. B* **43** 7231–7242
[25] Gonze X and Lee C 1997 *Phys. Rev. B* **55** 10355–10368
[26] Buckeridge J, Scanlon D O, Walsh A, Catlow C R A and Sokol A A 2013 *Phys. Rev. B* **87** 214304
[27] Mochizuki S 1982 *Phys. Status Solidi* **114** 189–199
[28] Kourouklis G A, Jayaraman A and Espinosa G P 1988 *Phys. Rev. B* **37** 4250–4253
[29] Clausen K, Hayes W, Macdonald J E, Osborn R, Schnabel P G, Hutchings M T and Magerl A 1987 *J. Chem. Soc. Faraday Trans. 2* **83** 1109–1112
[30] Nakajima A, Yoshihara A and Ishigame M 1994 *Phys. Rev. B* **50** 13297–13307
[31] Marabelli F and Wachter P 1987 *Phys. Rev. B* **36** 1238–1243

EXAFS simulations in Zn-doped $LiNbO_3$ based on defect calculations

Mário E G Valerio[1], Robert A Jackson[2], Frank G Bridges[3]

1. Department of Physics, Federal University of Sergipe, São Cristavão-SE, Brazil
2. School of Physical and Geographical Sciences, Keele University, Keele, Staffordshire ST5 5BG, UK
3. Department of Physics, University of California Santa Cruz, Santa Cruz CA 95064, USA

mvalerio@ufs.br

Abstract. Lithium niobate, $LiNbO_3$, is an important technological material with good electro-optic, acousto-optic, elasto-optic, piezoelectric and nonlinear properties. EXAFS on Zn-doped $LiNbO_3$ found strong evidences that Zn substitutes primarily at the Li site on highly doped samples. In this work the EXAFS results were revisited using a different approach where the models for simulating the EXAFS results were obtained from the output of defect calculations. The strategy uses the relaxed positions of the ions surrounding the dopants to generate a cluster from where the EXAFS oscillations can be calculated. The defect involves not only the Zn possible substitution at either Li or Nb sites but also the charge compensating defects, when needed. From previous defect modelling, a subset of defects was selected based on the energetics of the defect production in the $LiNbO_3$ lattice. From them, all possible clusters were generated and the simulated EXAFS were computed. The simulated EXAFS were them compared to available EXAFS results in the literature. Based on this comparison different models could be proposed to explain the behaviour of Zn in the $LiNbO_3$ matrix.

1. Introduction and Motivation

Zn-doped lithium niobate, $LiNbO_3$ is an important material for photorefractive applications. $LiNbO_3$ has been modelled in both pure and undoped states ([1], [2], [3]), and the calculations predict that Zn will substitute on the Li and Nb sites, with the defect $3Zn_{Li} - Zn'''_{Nb}$ being the defect with the smallest solution energy, when considering the bound defects. The defect involving charge compensation via Li vacancies, $Zn_{Li} - V'_{Li}$, where found to have energy three times higher than the previous one. A key issue in the previous modelling results is the fact that the defects were modeled as a full defect including all components in a way that they were neutral. Two important conclusions were drawn from this approach: i – the contribution of the binding energy to the total solution energy was very important, and for the case of the $3Zn_{Li} - Zn'''_{Nb}$ it reduced the total energy to half of the initial unbound value, and ii - the final configuration of the lattice is highly influenced by the presence of the defects and its charge compensation accompanying defects.

EXAFS (Extended X-rays Absorption Fine Structure) is a very powerful technique to investigate the symmetry of the surrounding neighborhood of the target absorbing ion in a material. Recent EXAFS studies ([4], [5]) suggest that Zn occupies the Li site, and the model was based on assuming as a starting point the Zn ion inside of a perfect LiNbO3 lattice and fitting the simulated EXAFS to the experimental results by varying the coordination shell distances, the disorder parameter (Debye-Waller factor) among others.

In the present paper, an alternative and complementary way of viewing the EXAFS results is proposed based on the results of previous defect modelling and the $LiNbO_3$:Zn system was used as a test case for such approach.

2. Modelling background

The details of the defect modelling calculations were described in previous papers [1], [2], [3] and the potential sets and defect configurations for the Zn ions in the $LiNbO_3$ matrix were the ones described by the Araujo, RM at al. [3]. The EXAFS simulations were done using the FEFF6 code embodied in the IFEFFit package [6].

The EXAFS simulations were carried out using clusters of radii around 8Å that were extracted from the output of the defect modelling calculations. The clusters, thus, already include the lattice distortion around the Zn ions due to the defect generated in the LiNbO3 lattice and since both the dopant and the charge compensating defects were included in the defect calculations, the distortion represents more accurately the system. The defect modelling calculations were done at 0K and this temperature was chosen due to the fact that the experimental data from refs [4] and [5] were measured at low temperature.

On building up the cluster to use as the input for the EXAFS modelling special attention were taken in the case of the $3Zn_{Li} - Zn'''_{Nb}$ defect because there are 4 Zn ions in the cluster and all of them will absorb the incident X-rays in the energy range covered by the EXAFS measurements in the Zn K edge. So, 4 different clusters were built where each one of the Zn ions were set at the origin and considered the absorbing atom, while the other ones acts as scatterers. The final simulated EXAFS was taken to be the average of the four individual simulations. The Debye-Waller factor was set to $0.003Å^{-2}$ that is a typical value for low temperature measurements and is close to the experimental values fitted to the EXAFS data in refs [4] and [5].

3. Results and Discussions

Figure 1 shows the simulated Fourrier transform of the EXAFS oscillations (red curves) for the model where Zn substitutes for the Li site (figure 1(a)) and substitutes at the Nb site (figure 1(b)) but considering the perfect lattice site with no distortion. Figure 1 also reproduces for comparison the experimental result of $LiNbO_3$:Zn sample (blue curves) extracted from reference [5]. For all simulated curves, the k space used in the Fourier transform of the EXAFS oscillations is compatible with the ones used in the experiment, i.e., k in the range from 3.8 to $13.2Å^{-1}$.

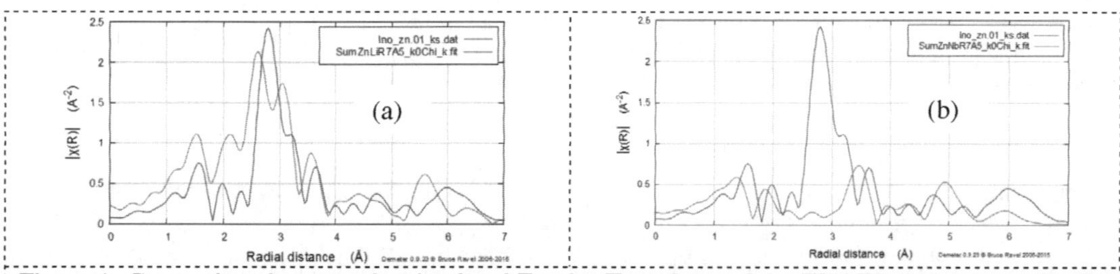

Figure 1: Comparison between the simulated Fourier Transform of the EXAFS (red curves) for Zn at a Li site (a) and Zn at Nb site (b) with the experimental measurments extracted from reference [5]. The results were expressed using k^2 weighting.

Results in figure 1 indicated that none of the models can describe the experimental results without taking into account distortion of the lattice. This is expected since Zn is a divalent metal and upon substituting either Li or Nb, the lattice will need to accommodate not only the differences on chemical behavior of the dopant in comparison to the host ions but also the charge compensating defects that will be needed. The computer modelling can be used to check how important these factors are. It is possible to consider first the effect of distortion assuming that the concentration of the dopants are low and there would be possible to have the charge compensating defect well apart from the dopant. Figures 1 indicated that the Zn substituting at the Li site seems to be closer to the experimental data than the other possibility and this was consider as a test.

Figure 2(a) shows the Fourier transform of the EXAFS simulations obtained for the Zn_{Li} relaxed configuration (red curve) as compared to the experimental data (blue curve). It is possible to see that the lattice distortion did improve a little bit the model, as compared to the experimental results, but the simulation is still quite far from the measurement. The next step was the inclusion of the charge compensation that, as discussed earlier, is important in the case of $LiNbO_3$ defects. The first attempt would be to include the simplest possible charge compensating that is a Li vacancy, although the defect simulations presented in ref [2] did show that this is not the defect with the lower solution energy. One can argue that since the concentration of Zn in the samples measured in ref [5] is low, the Zn ions should be far apart from each other and the neutral defect that would keep the Zn isolated is the $Zn_{Li} - V'_{Li}$ one. This can be tested by taking the relaxed lattice structure obtaining for the defect simulations of the $Zn_{Li} - V'_{Li}$ one and computing the EXAFS for that model, and the result of such procedure is shown in figure 2(b). The simulated EXAFS was further improved with the main peak, around 3Å showing a better agreement between the simulated and experimental curves, but the first peaks between 1 and 2 Å, due to the first shell of neighbors, typically O^{2-} ions, and the third group of peaks, around 6Å are not described by this latter model.

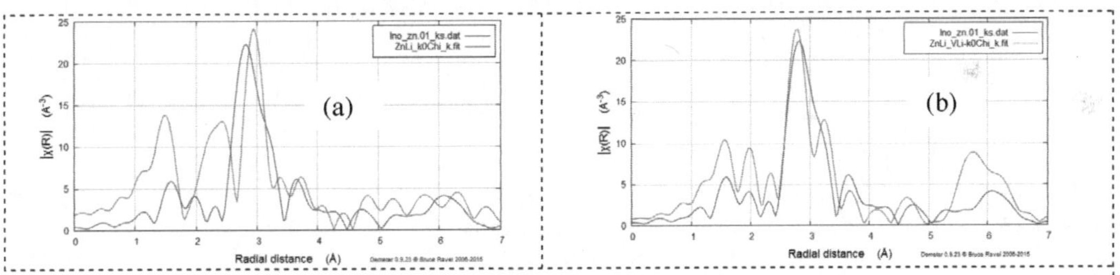

Figure 2: Comparison between the simulted Fourier Transform of the EXAFS (red curves) for the relaxed Zn_{Li} defect (a) and $Zn_{Li} - V'_{Li}$ relaxed defect (b) with the experimental measurments extracted from reference [5]. The results were expressed using k^2 weighting.

The next step is to test if the defect predicted by the defect modelling can improve the description of the experimental results. The defect predicted to have the smallest solution energy is the self-compensating defect $3Zn_{Li} - Zn'''_{Nb}$ that involves 4 Zn ions in the cluster that have to be treated to simulate the EXAFS, as explained before. The result is shown in figures 3 and 4, where figure 4 shows the contribution of the individual Zn ions in the cluster and figure 4 the averaged simulated EXAFS.

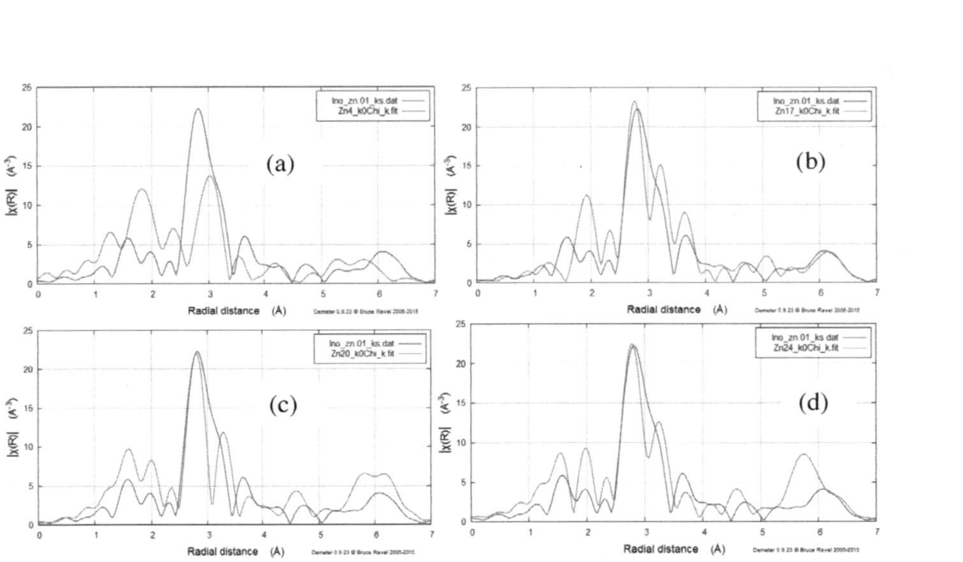

Figure 3: Comparison of the experimental measurments extracted from reference [5] with the simulted Fourier Transform of the EXAFS (red curves) for individual Zn ions in the relaxed $3Zn_{Li} - Zn'''_{Nb}$ defect: (a) Zn'''_{Nb} and (b), (c) and (d) Zn_{Li}. The results were expressed using k^2 weighting.

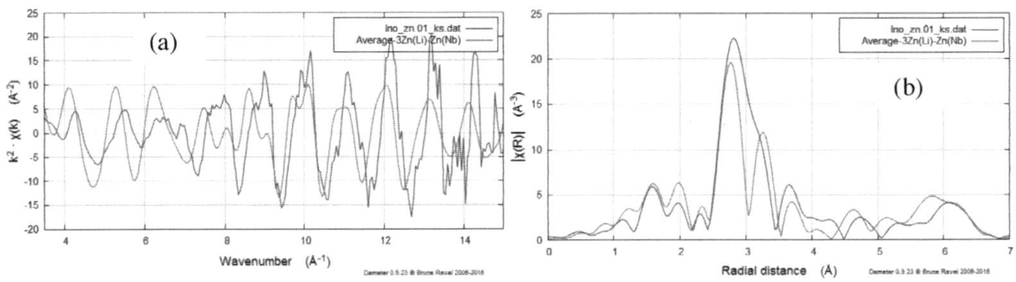

Figure 4: Comparison of the experimental measurments (blue curves) extracted from reference [5] with the simulted EXAFS oscilations (a) and Fourier Transform of the EXAFS (b) (red curves) for the relaxed $3Zn_{Li} - Zn'''_{Nb}$ defect. The results were expressed using k^2 weighting.

Figures 3 and 4 shows that the simulated EXAFS obtained for the relaxed $LiNbO_3$ lattice surrounding the $3Zn_{Li} - Zn'''_{Nb}$ defect gave a better description of the experimental data. The relative intensity of the peaks in the Fourier transform curve are much closer to the experimental results. It is interesting to notice that the position of the peaks due to different shell of neighbours (group of neighbouring ions around a same average distance from the target Zn atom) are reasonably well described by the configuration of the lattice after relaxation around the $3Zn_{Li} - Zn'''_{Nb}$ defect, although the EXAFS obtaining from the contributions of the individual Zn ions are not as good as the averaged EXAFS curves.

4. Conclusions

In conclusion the combination of defect modelling with the EXAFS simulations can be a powerful tool to help the interpretation of EXFAS data mainly in quite complex systems where a simple defect or a simple initial configuration of the lattice around the target absorbing ions cannot be easily devised.

The output of the defect simulations, that takes the relaxation of the lattice around the defect into account, can be used as an initial guess for the configuration of the lattice around the ion being probed. It is important to stress that this is not an alternative to the usual way of analyzing and interpreting EXAFS data, but a different and maybe a much easier way to interpret EXAFS data mainly for the case of dopants.

5. Acknowledgements

The authors are grateful to Prof. A V Chadwick and Prof. L Kovács for many fruitful discussions. This work was partially supported by CNPq (Conselho Nacional de Desenvolvimento Científico e Tecnológico) and FAPITEC-SE (Fundação de Apoio à Pesquisa e à Inovação Tecnológica do Estado de Sergipe).

References

[1] Jackson R A and Valerio M E G 2005 J. Phys.: Condens. Matter 17 837
[2] Araujo R M, Lengyel K, Jackson R A, Kovacs L and Valerio M E G 2007 J. Phys.: Condens. Matter 19 046211
[3] Araujo R M, Jackson R A and Valerio M E G 2008 J. Phys.: Condens. Matter 20 035201
[4] Bridges F, Castillo-Torres J, Car B, Medling S, and Kozina m 2012 PHYSICAL REVIEW B 85, 064107
[5] Bridges F, Mackeen C and Kovács L 2016 PHYSICAL REVIEW B 00, 004100 (accepted for publication)
[6] M. Newville, *J. Synchrotron Rad.* **8**, pp 322--324 (2001)
[7] B. Ravel and M. Newville, *J. Synchrotron Rad.* **12**, pp 537--541 (2005).

IOP Conf. Series: Materials Science and Engineering **169** (2017) 012004 doi:10.1088/1757-899X/169/1/012004

Computer modelling of double doped $SrAl_2O_4$ for phosphor applications

R A Jackson, L A Kavanagh and R A Snelgrove

Lennard-Jones Laboratories, School of Physical and Geographical Sciences, Keele University, Keele, Staffordshire, UK

r.a.jackson@keele.ac.uk

Abstract. This paper describes a modelling study of $SrAl_2O_4$, which has applications as a phosphor material when doped with Eu^{2+} and Dy^{3+} ions. The procedure for modelling the pure and doped material is described and then results are presented for the single and double doped material. Solution energies are calculated and used to deduce dopant location, and mean field calculations are used to predict the effect of doping on crystal lattice parameter. Possible charge compensation mechanisms for Dy^{3+} ions substituting at a Sr^{2+} site are discussed.

1. Introduction

$SrAl_2O_4$, when doped with Eu^{2+} and Dy^{3+} ions, is a well-known phosphor material, used widely in applications like emergency signs, which utilise its 'glow in the dark' properties [1]. Most experimental papers on this material suggest that doping only occurs at the Sr^{2+} site, but they do not discuss how the charge is compensated if a Dy^{3+} ion substitutes there. This paper addresses this specific question, and looks at how the structure changes in the presence of dopant ions. The relative energetics of single and double doping is also considered.

Computer modelling has been widely used to model single doped materials with optical applications [2,3], but there are relatively few examples of studies of double doped materials, except for a recent study of co-doping in $LiNbO_3$ [4]. An issue of interest is whether double doping can be an energetically favourable process when compared with single doping.

2. Methodology

2.1 Perfect lattice calculations. Calculations were performed using the GULP code [5]. Pure $SrAl_2O_4$ in its monoclinic phase was modelled using potentials derived by Catlow [6]. The starting experimental structure was obtained by Avdeev et al [7]. Table 1 shows the agreement between experimental and calculated structures.

Table 1: Comparison of experimental and calculated structure of $SrAl_2O_4$

	Experimental [7]	Calculated	% difference
A	8.44365	8.49801	0.64
B	8.82245	9.03433	2.40
C	5.15964	5.25031	1.76
$\alpha=\gamma$	90.0	90.0	0.0
β	93.411	92.425	-0.99

It is noted that the agreement is not perfect, but a compromise was made in that by using these potentials, other relevant oxide phases could be modelled so that solution energies could be calculated in section 3. However the agreement is good enough to proceed with defect calculations.

2.2 Defect formation and solution energy calculations. Defect formation energies were calculated using GULP [5], employing Mott-Littleton methodology [8]. Solution energies give the total energy involved in the solution process. For example, if a Eu^{2+} ion is doped at the Sr^{2+} site, the following reaction might occur:

$$EuO + Sr_{Sr} \rightarrow Eu_{Sr} + SrO$$

In this case, the solution energy for the reaction is given by:

$$E_{sol} = [E\ (Eu_{Sr}) + E_{latt}\ (SrO)] - [E_{latt}\ (EuO)]$$

Here, E (Eu_{Sr}) is the energy involved in doping a Eu ion at a Sr site, and E_{latt} (EuO) is the lattice energy of EuO. More complex solution schemes can be found in previous papers that use this approach [2-4]. It is noted that the potentials for the interactions of Eu^{2+} and Dy3+ ions with O^{2-} ions were obtained by fitting to the structures of EuO and Dy_2O_3 respectively, as discussed in [9].

2.3 Mean field calculations. Mean field calculations are perfect lattice calculations in which the occupancy of an ion/ions in the structure is changed, being replaced progressively by a dopant ion. In the case of doping $SrAl_2O_4$ with Eu^{2+} ions, a Sr site is chosen and Eu added at that site, corresponding to the structure $Sr_{1-x}Eu_xAl_2O_4$. The lattice parameters and properties can then be calculated as a function of dopant concentration. If the dopant is of greater charge than the doping site, the charge can be compensated by increasing the charge of the oxygen, which corresponds to adding oxygen interstitials. This is discussed further in section 3.3.

3. Results and Discussion

3.1 Defect formation energies. First, defect formation energies were calculated. In table 2 the energies are given, and it is noted that the value for the oxygen Frenkel energy was transferred from the Master's thesis of Marcos Rezende [9], and is there for comparison only.

Table 2: defect formation energies

	Energy/ eV
Sr vacancy	19.53
Al vacancy	59.26
O vacancy	25.16
E (latt) $SrAl_2O_4$	-194.4
E (latt) SrO	-33.97
Schottky (per ion)	6.325
SrO pseudo-Schottky (per ion)	5.330
O Frenkel (per ion) [9]	5.180

From the table it is clear that the lowest energy defect schemes involve SrO pseudo-Schottky and O Frenkel defects. In the calculations of solution energy charge compensation by Sr vacancies is assumed where needed; O interstitial formation will be considered in a later paper.

3.2 Solution energies for Eu^{2+} and Dy^{3+} doping. Solution energies were calculated for a number of possible scenarios for doping, and are described below.

3.2.1 Eu^{2+} doping at a Sr^{2+} site. For the simple case of doping a Eu^{2+} ion at a Sr^{2+} site, the equation given in section 2.2 applies. The calculated solution energy is 0.06 eV per Eu^{2+} ion, which suggests that solution is likely with little energy input. However, the issue of Dy^{3+} doping is more complex, and is considered next.

3.2.2 Dy^{3+} doping at the Al^{3+} site. If the Dy^{3+} ion dopes at an Al^{3+} site, the solution scheme may be assumed to be:

$$Dy_2O_3 + 2Al_{Al} \rightarrow 2Dy_{Al} + Al_2O_3$$

In this case the solution energy is calculated to be 1.72 eV per Dy^{3+} ion, which is still relatively low, and suggests that solution at this site is energetically feasible.

3.2.3 *Dy^{3+} doping at the Sr^{2+} with Sr^{2+} vacancy compensation.* The experimental papers tend to assume doping occurs at the Sr^{2+} site, but charge compensation is not discussed. If it occurs via formation of Sr^{2+} vacancies, the following scheme may apply:

$$Dy_2O_3 + 2Sr_{Sr} \rightarrow 2Dy^{\bullet}_{Sr} + V''_{Sr} + 3SrO$$

In this case the solution energy is calculated to be 3.08 eV per Dy^{3+} ion, which is considerably higher than for solution at the Al^{3+} site. The next step was to investigate whether co-doping with Eu^{2+} and Dy^{3+} would be energetically more favourable.

3.2.4 *Co-doping of Eu^{2+} and Dy^{3+} at Sr^{2+} sites with Sr^{2+} vacancy compensation.* The scheme below assumes that the charge from the substitution of two Dy^{3+} ions at Sr^{2+} sites is compensated by formation of a Sr^{2+} vacancy.

$$Dy_2O_3 + EuO + 4Sr_{Sr} \rightarrow Eu_{Sr} + 2Dy^{\bullet}_{Sr} + V''_{Sr} + 4SrO$$

The solution energy per dopant ion is calculated to be 2.08 eV, showing that co-doping can reduce the energy required in the doping process. However it is still quite a high energy, suggesting that alternative charge compensation schemes should be considered.

3.3 *Mean field calculations on M^{3+} doped $SrAl_2O_4$.* In order to investigate alternative charge compensation mechanisms, mean field calculations were carried out on M^{3+} doped $SrAl_2O_4$, comparing with results from an experimental study of Ce^{3+}, Pr^{3+} and Tb^{3+} doped $SrAl_2O_4$ [10]. In this paper, the lattice parameter of $SrAl_2O_4$ was measured when doped by the rare earth ions mentioned above, and the results for the 'a' parameter are given in the table below:

Table 3: Lattice parameter measurements of pure and doped $SrAl_2O_4$

Material	a/Å
Pure $SrAl_2O_4$	8.477
$SrAl_2O_4$: Ce^{3+}	8.431

$SrAl_2O_4$: Pr^{3+}	8.432
$SrAl_2O_4$: Tb^{3+}	8.441

Mean field calculations were carried out to see if it was possible to reproduce the reduction in lattice parameter observed in these measurements. The simplest charge compensation scheme, involving doping at the Al^{3+} site, resulted in lattice expansion. As an alternative, doping at the Sr^{2+} site was considered with charge compensation from an effective increase in the number of O^{2-} ions, leading to the structure $Sr_{1-x}M_xAl_2O_{4+x/2}$.

Figure 1 below shows the 'a' lattice parameter of $SrAl_2O_4$ measured as a function of dopant concentration, x (equivalent to M / %).

Figure 1: Mean field calculation of $SrAl_2O_4$ lattice parameter as a function of dopant concentration.

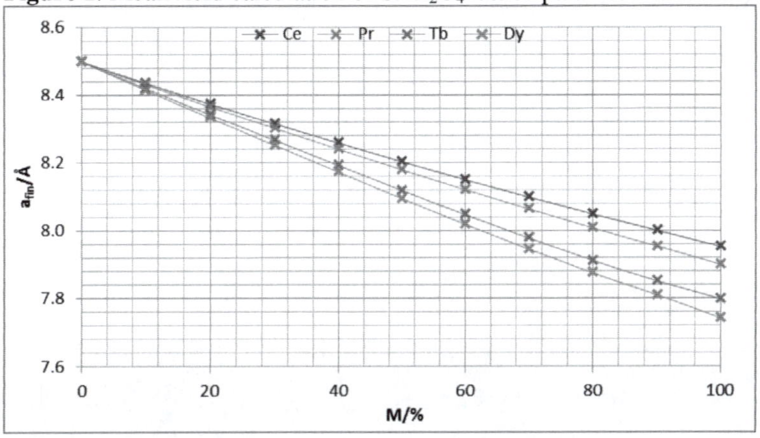

The calculations are shown to reproduce qualitatively the lattice contraction, which suggests that a charge compensation scheme based on O^{2-} interstitials should be considered. This will be the subject of a future publication.

4. Conclusions

The calculations presented in this paper have demonstrated the importance of understanding the charge compensation mechanisms associated with doping of materials, and has demonstrated how relatively simple solution energy calculations can help to achieve this. Further calculations are in progress to extend understanding of this process.

References

[1] Botterman J, Joos JJ and Smet PF 2014 *Physical Review B* **90** 085147
[2] Rezende MVS, Amaral JB, Valerio MEG and Jackson RA 2015, *Optical Materials* **48** 105
[3] Mujaji M, Burrows J and Jackson RA 2014 *Journal of Luminescence* **151** 106
[4] Araujo RM, Valerio MEG and Jackson RA 2014 *Proc. R. Soc. A* **470** 20140406
[5] Gale JD 1997 *J. Chem. Soc. Faraday Trans.* **93** 629
[6] http://www.ucl.ac.uk/klmc/Potentials/Library/catlow.lib
[7] Avdeev M, Yakovlev S, Yaremchenko AA and Kharton VV 2007 *J. Solid State Chem.* 180 3535
[8] Mott NF, Littleton MJ. 1938 *Trans. Faraday Soc.* **34** 485

[9] Rezende M 2008 MSc thesis, Federal University of Sergipe, Brazil
[10] Fu Z, Zhou S and Zhang S 2005 *J. Phys. Chem. B* **109** 14396

IOP Conf. Series: Materials Science and Engineering **169** (2017) 012005 doi:10.1088/1757-899X/169/1/012005

Changes in the optical absorption induced by sequential exposition to short- and long-wavelength radiation in the BTO:Al crystal

S M Shandarov [1], V G Dyu[1], M G Kisteneva[1], E S Khudyakova[1], S V Smirnov[1], A S Akrestina[1] and Yu F Kargin[2]

[1]State University of Control System and Radioelectronics, Tomsk, Russia
[2] A.A. Baikov Institute of Metallurgy and Materials Science RAS, Moscow, Russia

E-mail: stanislavshandarov@gmail.com[1]

Abstract. Modifications of the spectral dependences of the optical absorption induced in the $Bi_{12}TiO_{20}$:Al crystal as a result of sequential exposition to cw laser radiation first with the wavelength $\lambda_g = 532$ nm and then with the longer wavelength $\lambda_{l,n} = 588, 633, 655, 658, 663, 700, 780, 871$, or 1064 nm are investigated. We revealed that after the short-wavelength exposition to radiation with $\lambda_g = 532$ nm, the optical absorption in the crystal increases, and in the range 470–1000 nm, yields the spectrum whose form is independent of a prehistory. The subsequent exposition to longer-wavelength radiation leads to bleaching of the crystal in the examined spectral range. A maximum diminishing of the optical absorption in the crystal is observed upon exposure to radiation with the wavelength $\lambda_{l,5} = 663$ nm. To describe the experimentally observed reversible changes in the optical absorption spectrum in the $Bi_{12}TiO_{20}$:Al we use the impurity absorption model that takes into account the photoinduced transitions between two metastable states of a deep defect center leading to the change of its position in the crystal lattice under conditions of strong lattice relaxation.

1. Introduction

$Bi_{12}TiO_{20}$ (BTO) crystals belonging to the sillenite family are promising for application in dynamic holography based on contradirectional interaction of light waves in reflection geometry [1-4]. This is due to high values of the space charge field of reflection holograms formed in the BTO crystals due to considerable concentration of photoactive defect centers inherent in the crystals through which the photoinduced redistribution of the electric charge occurs [5, 6]. The special feature of the BTO crystals is the strong photochromic effect also observed at room temperatures; it consists in reversible changes of the optical absorption induced by exposition to visible radiation [7, 8]. The main reason for the increase in the optical absorption in the BTO crystals is considered to be the capture of photoexcited charge carriers (electrons) from deep donor centers to less deep traps characterized by the photoionization cross section larger than in the initial state [9 – 11]. Such process in which the nonequilibrium charge carriers participate can lead to an increase in the photorefractive sensitivity of the BTO crystals in the near-IR range of the spectrum [12, 13]. Another reason for the increased absorption in the sillenite crystals that is not associated with electron photoexcitation to the conduction band can be generation of defects for which intracenter transitions occur as a result of exposition to light [11]. The intracenter character of transitions in the BTO:Al crystal in the wavelength range from 500 to

900 nm characterized by a resonant dependence of the photo- and thermoinduced changes in the light absorption was suggested in [14]. In addition, third reason for changes in absorption may be the optically induced transitions between two metastable states of the deep defect centers in the crystal, which are accompanied by considerable reconstruction of such centers manifested via changes in their positions in the lattice [15].

In this paper we experimentally investigate the changes in spectral dependences of the optical absorption in the $Bi_{12}TiO_{20}$:Al crystal caused by sequential exposition to cw laser radiation to the shortest wavelength $\lambda_g = 532$ nm in the first stage followed by exposition to the longer wavelength $\lambda_{l,n}$ taking in different experiments value of 588, 633, 655, 658, 663, 780, 871, or 1064 nm, respectively. To describe the observed reversible changes the model taking into account strong relaxation of the lattice during photoinduced transitions between two metastable states of a deep defect center, in addition to the impurity absorption model included both the photoexcitation of electron from deep centers to the conduction band and the intracenter transitions [11], is considered.

2. Experimental method and results

The examined $Bi_{12}TiO_{20}$:Al sample was grown by the TSSG method and had the thickness $d = 6.6$ mm along the [100] crystallographic direction. Its transmission spectra were registered for the initial crystal state and after each exposition. All experiments were performed at room temperature.

The experimental investigations performed previously in [16] demonstrated that exposition to short-wavelength radiation with the wavelength $\lambda_g = 532$ nm caused an increase in the optical absorption coefficient $k_g(\lambda)$ in the BTO:Al crystal and resulted in the spectrum whose shape in the wavelength range 470–1000 nm was independent of the initial crystal state. We established that the subsequent exposition to a longer-wavelength radiation with $\lambda_{l,n}$ changing from 588 to 1064 nm led to the enhanced transmittance of the BTO:Al crystal in the entire examined spectral range with absorption coefficients $k_{l,n}(\lambda)$, which reaches saturation depending on the employed wavelength $\lambda_{l,n}$.

Even though the dependences $k_g(\lambda)$ and $k_{l,n}(\lambda)$ were monotonic [16], the spectral dependences of the changes in absorption $\Delta k_n = k_g(\lambda) - k_{l,n}(\lambda)$ demonstrated a resonant character. Figure 1 shows the spectral dependences characterising the maximum changes in the optical absorption for the $Bi_{12}TiO_{20}$:Al crystal with enhanced transmittance, preliminary illuminated by green light ($\lambda_g = 532$ nm), reached by its subsequent illumination with longer-wavelength radiation at $\lambda_{l,n}$= 663 (curve 1), 655 (curve 2), 633 (curve 3), 780 (curve 4), 871 (curve 5), or 1064 nm (curve 6).

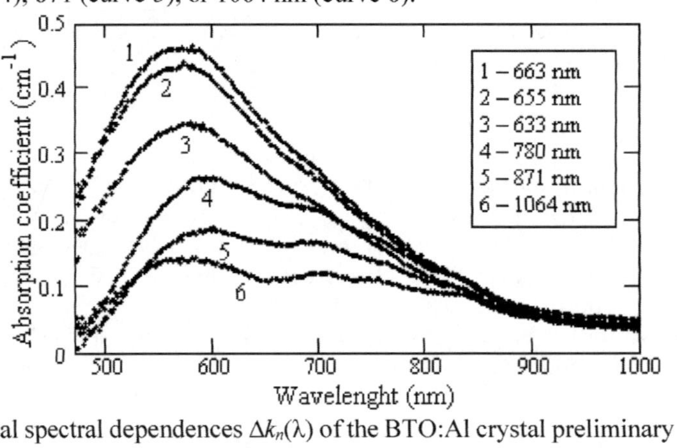

Figure 1. Experimental spectral dependences $\Delta k_n(\lambda)$ of the BTO:Al crystal preliminary illuminated with laser radiation $\lambda_g = 532$ nm observed after subsequent illumination with $\lambda_{l,n}$= 663 (1), 655 (2), 633 (3), 780 (4), 871 (5), or 1064 nm (6) until saturation.

It should be noted that the maximum decrease of the optical absorption in the $Bi_{12}TiO_{20}$:Al crystal was observed after exposition at the wavelength $\lambda_{l,5} = 663$ nm. The changes in the optical absorption during enhanced-diminished transmittance cycles were reversible and well reproduced experimentally.

3. Theoretical model

As is known [15, 17], localized optically induced changes in the electronic configuration of the deep defect centers of some types in semiconductor and dielectric materials are accompanied by considerable reconstruction of such centers manifested via changes in their positions in the lattice. This effect called the photoinduced lattice relaxation [17] leads to the possibility of photocontrollable change of the properties of crystals comprising such centers and of their application as recording media [15]. The configuration coordinate diagram [17] best described the change in the geometry of the defect caused by the change in its electronic state.

Let us consider the deep defect center that can be in two metastable states A and C with the potential energy minima shifted along the configuration coordinate Q (figure 2). The parabolas $E_{d1}(Q)$ and $E_{d2}(Q)$, described by the well-known relations [18]

$$E_{d1}(Q) = a_1 (Q - Q_1)^2, \tag{1}$$

$$E_{d2}(Q) = E_{02} + b_2 (Q - Q_2)^2, \tag{2}$$

characterize the energies of the ground (1) and excited (2) states of the centers with the parameters a_1 and b_2 of the parabolic functions that differ from each other taking into account the thermal excitation energy E_{02}.

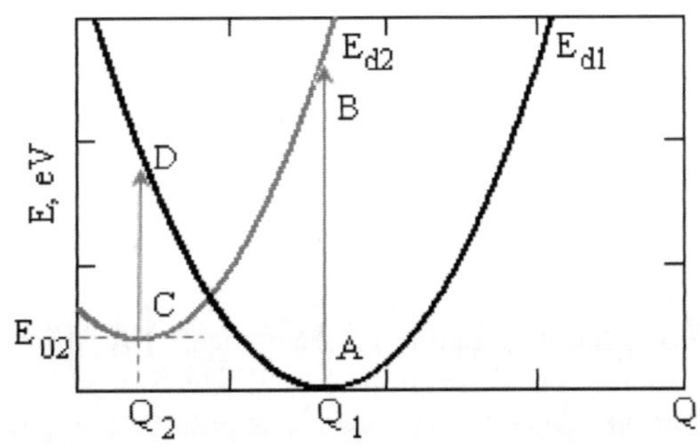

Figure 2. Configuration coordinate diagram of a defect center.

For vertical transitions that can be induced only by light radiation, the configuration coordinates Q_{v1} and Q_{v2} corresponding to the energy $\hbar\omega_{12}$ of optical transitions from the ground to the excited state

$$\hbar\omega_{12} = E_{d2}(Q_{v1}) - E_{d1}(Q_{v1}), \tag{3}$$

and to the energy $\hbar\omega_{21}$ of optical transitions back to the ground state

$$\hbar\omega_{21} = E_{d1}(Q_{v2}) - E_{d2}(Q_{v2}), \tag{4}$$

can be obtained from equations (1)–(4) in the following form:

$$Q_{v1}(\hbar\omega) = \frac{a_1 Q_1 - b_2 Q_2}{a_1 - b_2} - \left[\left(\frac{a_1 Q_1 - b_2 Q_2}{a_1 - b_2} \right)^2 - \frac{E_{02} - \hbar\omega + a_1 Q_1 - b_2 Q_2}{a_1 - b_2} \right]^{1/2}, \tag{5}$$

$$Q_{v2}(\hbar\omega) = \frac{a_1 Q_1 - b_2 Q_2}{a_1 - b_2} - \left[\left(\frac{a_1 Q_1 - b_2 Q_2}{a_1 - b_2} \right)^2 - \frac{E_{02} + \hbar\omega + a_1 Q_1 - b_2 Q_2}{a_1 - b_2} \right]^{1/2}. \tag{6}$$

Assuming that the energy distributions for the defect center concentrations in the ground state, $N_{d1}(E_{d1})$, and in the excited state, $N_{d2}(E_{d2})$, are described by the Boltzmann distributions, we find the absorption coefficients $k_{d1}(\omega)$ and $k_{d2}(\omega)$ caused respectively by the optical transitions from the ground state to the excited state and by the reverse transitions, as

$$k_{d1}(\omega) = \hbar\omega S_{d1} N_{d10} \exp\left(-\frac{a_1 (Q_{v1} - Q_1)^2}{k_B T}\right), \quad (7)$$

$$k_{d2}(\omega) = \hbar\omega S_{d2} N_{d20} \exp\left(-\frac{b_1 (Q_{v2} - Q_2)^2}{k_B T}\right), \quad (8)$$

where N_{d10} and N_{d20} are particle concentrations in the ground state for $Q = Q_1$ and in the excited metastable state for $Q = Q_2$, respectively; S_{d1} and S_{d1} are the photoabsorption cross sections for the ground and excited state, respectively; k_B is the Boltzmann constant, and T is the absolute temperature.

Thus, the optical absorption in the crystal caused by the examined deep defect center is spectrally dependent, and its absorption coefficient $k_d(\omega)$ is determined by the sum

$$k_d(\omega) = k_{d1}(\omega) + k_{d2}(\omega), \quad (9)$$

depending, according to equations (7), (8) and (5), (6), on the concentrations of the centers in the ground and excited states, the thermal excitation energy E_{02}, differences between their photoabsorption cross sections and the energy parameters a_1 and b_2, and the shift between the ground and excited states on the configuration coordinate $Q_1 - Q_2$. The optical transitions induced by exposition of the crystal to radiation with a certain frequency ω_n lead to redistribution of the center concentration between these states. An analysis demonstrates that the equilibrium condition defined by equality of the numbers of direct and reverse transitions is established when the condition

$$k_{d1}(\omega_n) = k_{d2}(\omega_n) \quad (10)$$

is satisfied.

4. Approximation of the spectral dependences of the optical absorption

The spectral dependences of the optical absorption in the BTO:Al crystal can be approximated by the impurity absorption model that takes into account the contribution of electrons photoexcited from deep donor centers to the conduction band and the intracenter transitions [11] supplemented by the model of the defect center interacting with the lattice that is based on the application of the configuration coordinate diagram considered above. Below we restrict ourselves by an analysis of three experimental dependences from [16] shown by open circles in figure 3. Dependence 1 in this figure shows the absorption spectrum for the BTO:Al crystal exposed in the initial stage of each experiment to laser radiation with the wavelength λ_g =532 nm. Spectral dependences 2 and 3 shown in figure 3 illustrate the subsequent exposition of the crystal to continuous laser radiation with the wavelength $\lambda_{l,5}$ = 663 nm and $\lambda_{l,7}$ = 780 nm, respectively, in the second stage of the experiments.

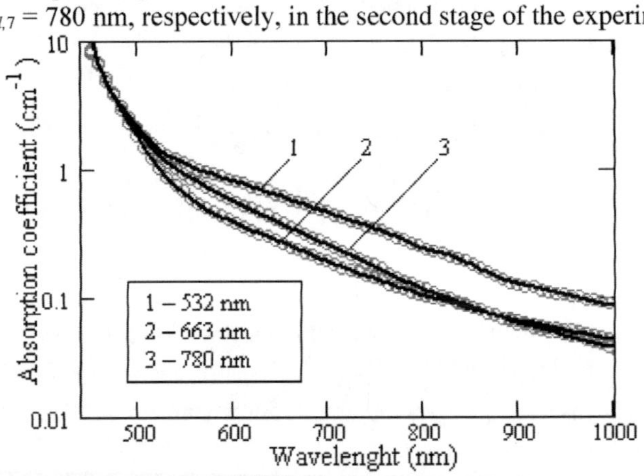

Figure 3. Experimental (open circles) and calculated (solid curves) spectral dependences of the optical absorption coefficient for the BTO:Al crystal exposed first to laser radiation with λ_g = 532 nm (1) and then to radiation with $\lambda_{l,n}$= 663 (2), 780 (3), respectively, until saturation.

The approximation of the experimental spectral dependences of the optical absorption in the BTO:Al crystal yielded the following values of the model parameters for the absorption caused by deep defect centers and described by equations (7) and (8): $a_1 = 5.426$ eV, $b_2 = 6.480$ eV, $E_{02} = 0.103$ eV; $Q_1 = 0.8$, $Q_2 = 0.2$; $S_{d1} = 3.94 \cdot 10^{-5}$ cm^2·eV^{-1} and $S_{d2} = 1.0 \cdot 10^{-4}$ cm^2·eV^{-1}. The parameters N_{d10} and N_{d20}, depending on the illuminating radiation wavelength ensuring the fulfillment of balance condition (10) for transitions between the deep center states, are presented in table 1 for laser radiation with three wavelengths and for nonmonochromatic radiation of a light-emitting diode with the average wavelength $\lambda_{l,1} = 588$ nm. For nonmonochromatic radiation, the experimental technique was analogous to that used in [16]. In the first stage, the BTO:Al sample was irradiated by laser radiation with the wavelength $\lambda_g = 532$ nm that caused the diminished transmittance of the crystal; the subsequent sample exposition to incoherent radiation with $\lambda_{l,1} = 588$ nm and intensity $I_1 = 0.3$ mW/cm^2 in the second stage caused the enhanced transmittance of the crystal with saturation at $t_1 = 2760$ s.

To approximate satisfactorily the spectral dependences, three intracenter transitions with the Gaussian spectral characteristics and maxima at quantum energies of 1.49, 1.62, and 1.77 eV must be taken into account together with the electron photoexcitation to the conduction band from four donor centers with average ionization energies of 1.08, 1.56, 1.96, and 2.75 eV. The energy parameters of the defect centers in the BTO:Al crystal were close to those used in [11]. The approximated spectral dependences shown in figure 3 by solid curves demonstrate good agreement with the available experimental data.

Table 1. Parameters of the deep defect center model depending on the irradiated wavelength

λ, nm	N_{d10}, m^{-3}	N_{d20}, m^{-3}
532	$4.68 \cdot 10^{23}$	$7.33 \cdot 10^{23}$
588	$8.55 \cdot 10^{23}$	$3.45 \cdot 10^{23}$
663	$10.7 \cdot 10^{23}$	$1.33 \cdot 10^{23}$
780	$11.5 \cdot 10^{24}$	$0.50 \cdot 10^{23}$

An analysis of the model parameters used to approximate the experimental data demonstrated the following. The increased optical absorption induced in the BTO:Al crystal by exposition to short-wavelength radiation with the wavelength $\lambda_g = 532$ nm was due to the increased contribution of the intracenter transitions to the radiation absorption, photoinduced increase in the degree of occupation by electrons of the centers with ionization energies of 1.08 and 1.96 eV, and photoexcitation of deep defect centers and their transition from the ground state to the excited metastable state (see figure 2). The subsequent exposition to long-wavelength radiation led to the reduced contributions to the absorption for intracenter transitions and as the result of the light-induced depletion of occupation by electrons of the centers with ionization energies of 1.08 and 1.96 eV. In this case, the degree of occupation by electrons of the center with ionization energy of 1.56 eV increased, and the concentration of the deep defect centers in the excited state decreased due to their photoinduced transitions to the ground state.

Thus, the observed nonmonotonic dependence of the enhanced transmittance for the $Bi_{12}TiO_{20}$:Al crystal, having the diminished transmittance, on the wavelength $\lambda_{l,n}$ of radiation leading to the enhanced transmittance can be caused not only by quantitative changes of contributions to the impurity absorption described in [11], but also by photoinduced transitions of the deep defect center between the ground and excited states, accompanied by strong electron-phonon interaction with distortion of the crystal lattice. For energy of the radiation quantum with $\lambda_g = 532$ nm corresponding to the vertical transition from the state A (figure 2) with maximum concentration of the centers to the ground state, they are effectively translated to the excited state (AB transition), leading to the diminished transmittance of the crystal. The enhanced transmittance of the crystal will be maximum for radiation with energy of the quantum at which the CD transition occurs (in the examined case, at the wavelength $\lambda_l \approx 670$ nm).

5. Conclusions

Thus, illumination of the BTO:Al crystal with cw laser radiation at the wavelength λ_g = 532 nm yields maximum values of the optical absorption in the range 470–1000 nm and the spectral dependence $k_g(\lambda)$ independent of the initial crystal state. Subsequent illumination of the examined sample with longer-wavelength laser radiation at the wavelength $\lambda_{l,n}$ = 588, 633, 655, 658, 663, 700, 780, 871, or 1064 nm leads to its enhanced transmittance. The minimum values of $k_{l,n}(\lambda)$ were observed after illumination of the BTO:Al crystal by light at $\lambda_{l,5}$ = 663 nm. The changes in the optical absorption spectrum for the BTO:Al crystal are well described by the impurity absorption model, considering the photoexcitation of electrons from the deep centers to the conduction band and the intracenter transitions [11], complemented by the model of deep defect centers with two metastable states interacting with the lattice.

Acknowledgments

This work was supported in part by the State Assignment of the Ministry of Education and Science of the Russian Federation for 2015–2016 and R&D under the Project of State Assignment No. 3.878.2014/IK, and the RFBR (grant No. 16-29-14046-ofi_m).

References

[1] Kukhtarev N, Chen B S, Venkateswarlu P, Salamo G and Klein M 1993 *Opt. Commun.* **104** 23
[2] Kamshilin A A, Romashko R V and Kulchin Y N 2009 *J. Appl. Phys.* **105** 031101
[3] Kolegov A A, Shandarov S M, Simonova G V, Kabanova L A, Burimov N I, Shmakov S S, Bykov V I and Kargin Y F 2011 *Quant. Electron.* **41** 847
[4] Shandarov S M, Shmakov S S, Zuev P V, Burimov N I, Kargin Y F, Shepelevich V V, Ropot P I and Gudelev V G 2013 *J. Opt. Technol.* **80** 409
[5] Shandarov S M, Burimov N I, Kul'chin Y N, Romashko R V, Tolstik A L and Shepelevich V V 2008 *Quant. Electron.* 1059
[6] Shandarov S M, Shmakov S S, Burimov N I, Syuvaeva O S, Kargin Y F and Petrov V M 2012 *JETP Letters* **95** 618
[7] Mersch F, Buse K, Sauf W, Hesse H and Krätzig E 1993 *Phys. Status Solidi* A **140** 273
[8] Matusevich A, Tolstik A, Kisteneva M, Shandarov S, Matusevich V, Kiessling A and Kowarschik R 2008 *J. Appl. Phys.* B **92** 219
[9] Kobozev O V, Shandarov S M, Kamshilin A A and Prokofiev V V 1999 *J. Opt.* A **1** 442
[10] Marinova V, Lin S H, Sainov V, Gospodinov M and Hsu K Y 2003 *J. Opt.* A **5** S500
[11] Kisteneva M G, Khudyakova E S, Shandarov S M, Akrestina A S, Dyu V G and Kargin Y F 2015 *Quant. Electron.* **45** 685
[12] Odoulov S G, Shcherbin K V and Shumeljuk A N 1994 *J. Opt. Soc. Am.* B **11** 1780
[13] Dos Santos P V, Frejlich J and Carvalho J F 2005 *J. Appl. Phys.* B **81** 651
[14] Kisteneva M G, Shandarov S M, Akrestina A S, Popugaeva V V and Smirnov S V 2010 *Izv. Vyssh. Uchebn. Zaved. Fiz.* **9/3** 145
[15] Ryskin A I and Fedorov P P 1997 *Phys. Solid State* **39** 943
[16] Dyu V G, Kisteneva M G, Shandarov S M, Khudyakova E S, Smirnov S V and Kargin Y F 2015 *Phys. Procedia* **73** 131
[17] Shinozuka Y 1993 *Jpn. J. Appl. Phys.* **33** 4560
[18] Ridley B K 1982 *Quantum Processes in Semiconductors* (Oxford University Press) p 304

Study of point defects in as-drawn and irradiated Ge-doped optical fibers using cathodoluminescence

I. Reghioua[1], S. Girard[1], A. Alessi[1], D. Di Francesca[1], E. Marin[1], A. Morana[1], M. Fanetti[2], L. Martin-Samos[2], N. Richard[3], M. Raine[3], M. Valant[2], A. Boukenter[1], Y. Ouerdane[1]

1 Univ-Lyon, Laboratoire Hubert Curien, UMR CNRS 5516, Saint-Etienne, France
2 Materials Research Laboratory, University of Nova Gorica, Vipavska 11c 5270-Ajdovscina, Slovenija
3 CEA, DAM, DIF, F-91297, Arpajon, France

Imene.reghioua@univ-st-etienne.fr

Abstract. In the present paper, we report an experimental investigation of Ge-doped Optical Fibers (OFs) which were investigated by Cathodoluminescence (CL) measurements. We followed the evolution, under 10 keV electron exposure, of the emissions present in three different samples: the first one was the as-drawn fiber (pristine), the second one was irradiated with a CW UV laser at 244 nm and the last one was irradiated at the dose of 9 MGy (SiO_2) by γ-rays. Moreover, taking advantage of the employed experimental set-up, which allows to perform spatially-resolved (<1μm) CL measures, we were able to investigate the emission evolution in two differently doped zones of the fiber. Our data indicate that (i) the CL spectra of our three samples are dominated by the 400 nm emission band related to the Germanium Lone Pair Center (GLPC), (ii) the spatial distribution of this defect differs in the three fibers and (iii) the electron exposure decreases the GLPC concentration in all samples (pristine, UV and γ irradiated). A comparison between the CL and photoluminescence (PL) measurements shows comparable results.

1. Introduction

Recently, the different harsh environments such as those associated with nuclear power plants, space missions or various research facilities are considered for Optical Fibers (OFs) integration as parts of data transfer or sensing systems (temperature, strain, dose) [1]. Thanks to their high performances, silica based OFs have been shown to be promising candidates for operation in such severe conditions (different types of radiations, temperature…).

When OFs are exposed to either ionizing or non-ionizing radiations like γ-rays, X-rays or electrons, three main effects are induced: the first one is the Radiation Induced Attenuation (RIA), which degrades the transmitted signal via the absorbing point defects; the second one is the Radiation Induced Emission (RIE) that decreases the signal to noise ratio; the third one is the compaction phenomenon that can induce a refractive index change [1].

Among the diversity of the OFs, the Ge-doped ones are of great importance. They are used for the telecom-grade applications thanks to the Ge ability to increase the refractive index of the OF's cores [2]. Also, their photosensitivity property is used for Fiber Bragg Gratings (FBG) inscription [3]. However,

doping with Ge will induce Ge-related point defects that need to be well controlled under irradiation. Many previous investigations have dealt with the radiation effects on Ge-doped silica [4-7]. Among the Ge-related point defects, the Germanium Lone Pair Center (GLPC) plays a key role for understanding the radiation effects on germanosilicate OFs [8, 9]. This defect consists in twofold coordinated Ge atom owning a lone pair of electrons, and it is characterized by two absorption bands at ~5.1 and ~3.8 eV and two emission bands at ~4.2 and 3.1 eV [10]. Under irradiation, this defect is converted into other point defects such as Ge(1) and Ge(2) [8, 9]. The Ge(1) is formed by an unpaired electron that is trapped in a substitutional tetra-coordinated Ge atom [11, 12]. For Ge(2), a few models have been put forward: the first one in [11] defines the Ge(2) as a variant type of the Ge(1) center, while a second model [12] considers that Ge(2) is an ionized GLPC, a third model suggests that the Ge(2) is a GLPC converted into a threefold forward oriented Ge atom [13].
Cathodoluminescence (CL) is a pertinent technique in order to study the luminescent point defects in dielectric materials like silica, since it allows to perform spatially-resolved in-situ measurements by using both CL spectroscopy and CL imaging. Various studies have been performed before using CL for investigating point defects and electron radiation effects on silica, such as in ref [14-16], but few of them were devoted to Ge-doped OFs [17-20]. In this work, we employed a CL-based apparatus to investigate the effects of different radiations (γ-rays, UV, Electrons) on germanosilicate OFs.

2. Experimental details

2.1. EDX and CL set-up

A field emission Scanning Electron Microscopy (SEM) (JSM 7100f JEOL) equipped with Energy Dispersive X-ray (EDX) detector (X-Max 80, OXFORD) and CL detector (GATAN MONOCL4) has been used for both EDX and CL measurements.
The following results have been obtained using a beam energy of 10 keV and a probe current in the order of 10 nA. These conditions lead to a penetration depth of about 1.3 μm and a spatial resolution of ~1μm.
Concerning the CL measurements, the employed instrument works in the 300-750 nm spectral range, and it allows to perform two kinds of CL measurements: CL spectroscopy that is used for the identification of the emitting defects and CL imaging, which is used to investigate the spatial distribution of the defects. The panchromatic imaging is performed by collecting all the emitted light whereas the monochromatic imaging consists in isolating a narrow spectral range (11 nm) in order to single out the associated spatial distribution.
For the kinetics data, an imaging analysis has been used for the intensity calculation which has been already described in ref [20].

2.2. Photoluminescence set-up

In the present investigation, the Photoluminescence (PL) measurements were carried out with an Aramis (Jobin-Yvon) confocal system which employs a HeCd laser at 325 nm as excitation and a ×40 UV objective which allows a spatial resolution of ~1μm.

2.3. Samples description

The employed OFs are multimode with a cladding diameter of 125 μm and a core diameter of 62.5 μm. The samples were manufactured using MCVD (Modified Chemical Vapor Deposition) process and were provided by iXBlue photonic division. As it is shown in Fig.1, the core of these OFs has a 2 steps Ge-doping profile: the highest doping level, ~9 wt%, is situated in the inner part of the core, the second one, ~ 5 wt%, is situated in the outer part of the core.
Two samples of the same OF have been differently irradiated while the third one was left as a pristine reference. The first irradiated sample was exposed to γ-rays up to 9MGy (SiO_2) using the Brigitte ^{60}Co source of SCK-CEN. The second sample was irradiated by UV continuum laser at 244 nm with a spot of ~0.5 mm using ~110 mW of power, by moving the sample longitudinally with a speed of 0.13 cm/s. Using the relation employed in [21] the fluence of the laser radiation is ~ 21.5 J/cm^2.
The different samples have been cleaved and their transverse cross sections has been exposed to a normal incident electron beam associated to the CL measurements.

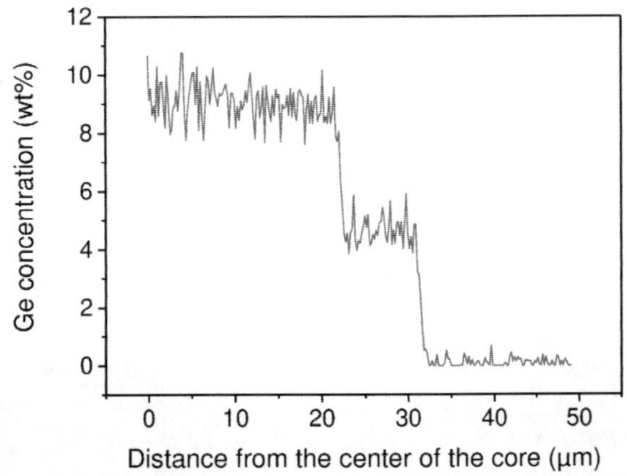

Figure 1. EDX measurements of the Ge concentration

3. Results and discussion

Fig. 2.a reports the CL panchromatic image of the transverse cross section of the pristine OF; it shows that the emitted light is mainly provided by the core part of the OF. The normalized CL spectra reported in the panel (b) of the same figure show the presence of an emission band centered at ~400 nm in the two Ge doped regions. This band is assigned to the GLPC point defects [10]. The pure silica cladding appears darker in the panchromatic image, which means that the intensity of the different intrinsic emitting defects present in this part is much lower compared with the one emitted by the core. Three bands are present is this cladding part, the first one at ~460 nm is attributed to the ODC (II) (oxygen deficient center) [10, 22, 23], the origin of the second band centered at 560 nm is still undetermined, since it could be either the second order of the emission of ODC (II) at 280 nm or a STE (Self-Trapped Exciton) center according to some papers in literature [16] or a sum of both of them. The last band at 650 nm is an emission band of the NBOHC (Non Bridging Oxygen Hole Center) [10, 22, 23]. The same CL spectra have been obtained for the two irradiated samples. These data indicate that the Ge doping induces additional luminescent defects in the studied spectral range.

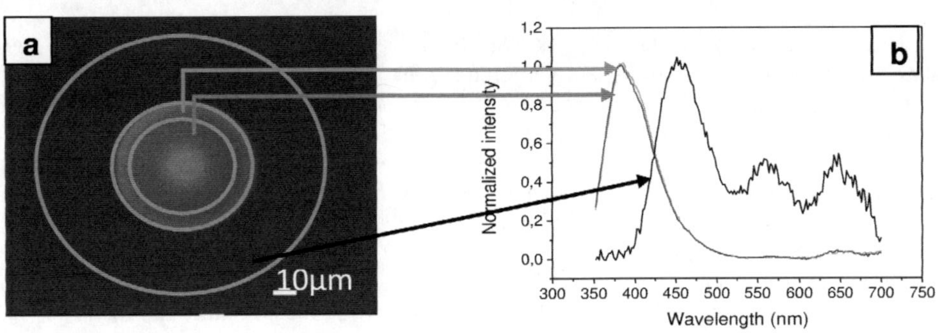

Figure 2. a: Panchromatic image of the pristine sample. b: CL spectroscopy of the different regions of the OF.

The Panchromatic images of the 9MGy γ irradiated fiber as well as the UV irradiated one are shown in Fig.3.a, b and Fig.3.c, d respectively before and after being exposed to 10 keV of electron radiation. Please note that different samples of the same OFs were used to perform the panchromatic images and the monochromatic image analysis. For panchromatic images, there are some differences in the used initial conditions for brightness and contrast, which explains the difference in absolute intensity between

IOP Conf. Series: Materials Science and Engineering **169** (2017) 012006 doi:10.1088/1757-899X/169/1/012006

the panchromatic images of the pristine, 9 MGy γ irradiated and UV irradiated samples before electron irradiation (Fig.2a, 3a and 3c). On contrary, these conditions were kept constant for the monochromatic images acquisition, which allows to directly compare the profiles presented in Fig.4b and 4c and in the study of the GLPC kinetics (Fig.5).

These data show that for both samples (9 MGy and UV irradiated) the electron irradiation caused a significant decrease of the luminescence in the core of the OFs. According to the previous reported spectra, this luminescence is mainly due to the GLPC. Moreover, before irradiating with electrons, the luminescence was more intense in the inner part of the core, where the Ge concentration is higher, whereas after irradiation, the outer part of the core is brighter. The same behavior was observed for the pristine sample [20].

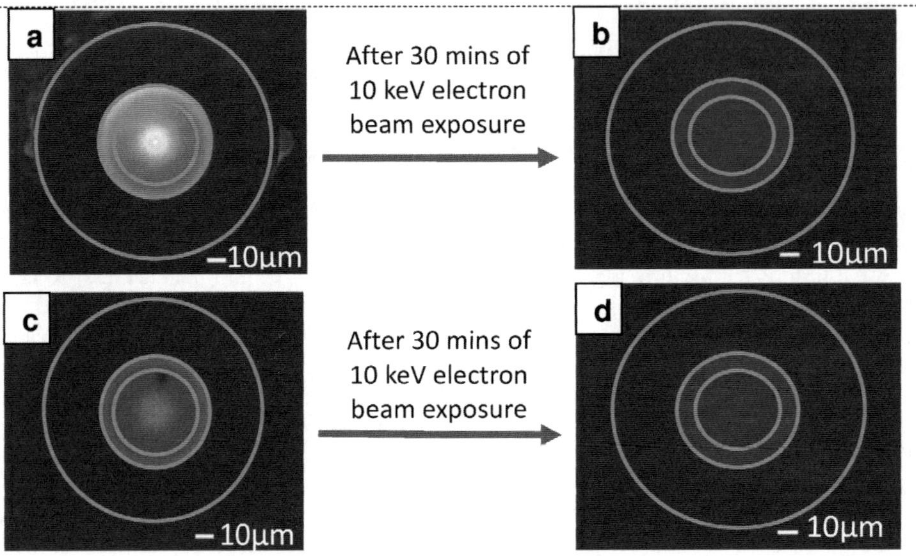

Figure 3. CL panchromatic images, a and b: the 9 MGy γ pre irradiated sample before and after electron irradiation respectively. C and d: the UV pre irradiated sample before and after electron irradiation respectively.

Concerning the pure silica part of the samples (cladding), the reported panchromatic images show no significant changes in the intrinsic luminescent defects in the investigated spectral domain. This could be due to the experimental conditions (adjustment of the image acquisition) which prevent us to detect the intrinsic pure silica luminescent defects compared to the Ge related ones.

Fig.4 shows the different radial profiles along the core radius of the different samples at 400 nm: for PL measurements (a), and for CL measurements before (b) and after (c) electron radiation. These profiles have been obtained by performing a mapping for PL measurements and by acquiring monochromatic images for the CL measurements. Fig.4.a and fig.4.b report the spatial distributions of the GLPC before irradiation with electrons, obtained with two different techniques. We can note that the ratio between the pristine and the 9 MGy γ irradiated samples is ~0.77 for CL measurements and ~0.63 for the PL ones, which means that the pre irradiation with γ-rays up to 9 MGy has induced a reduction of the GLPC signal of about 30%. Regarding the UV irradiated sample (fig.4.b), we note that according to the CL measurements, no significant change was induced in the GLPC intensity compared with the pristine sample. This could be due to the small accumulated dose deposited in the sample during irradiation.

After being exposed to 10 keV of electron beam for 30 mins, the three samples show a similar radial distribution of the GLPC luminescence (see Fig.4.c). This luminescence decreased significantly and its signal is higher in the lower Ge-doped region, confirming the previous panchromatic images reported in Fig.3.

IOP Conf. Series: Materials Science and Engineering **169** (2017) 012006 doi:10.1088/1757-899X/169/1/012006

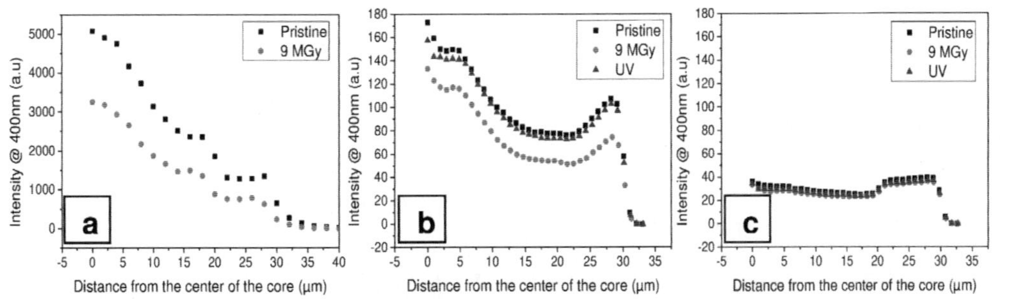

Figure 4. Radial distributions along the core radius of the luminescence at 400 nm a. PL measurements. b. CL measurements before electron exposure. c. CL measurements after 30 mins of electron exposure.

In order to clarify the Ge impact on the bleaching process of GLPC, an in-situ investigation under electron exposure in the different parts of the OF's core is reported in Fig.5. The GLPC kinetics has been obtained by acquiring a series of monochromatic images at 400 nm. We observe that the GLPC signals were bleached in both the inner and the outer regions of the core and reached a minimum level by the end of the irradiation. After 20 scans (~6 mins) the intensity of the signal at 400 nm has been decreased from ~150 to ~50 in the 9 wt% Ge-doped part, while it has been reduced from ~90 to ~40 in the 5 wt% Ge-doped zone. Also, the 9 MGy irradiated sample showed a lower GLPC concentration in the first recorded image (the first scan) confirming a bleaching effect induced by the γ pre irradiation. This bleaching behavior under radiation could be explained by a conversion process of the GLPC to other Ge-related centers [8]. After a sufficiently long electron irradiation, the GLPC kinetics reached equilibrium due to the radiation induced conversion and the back conversion of these centers.

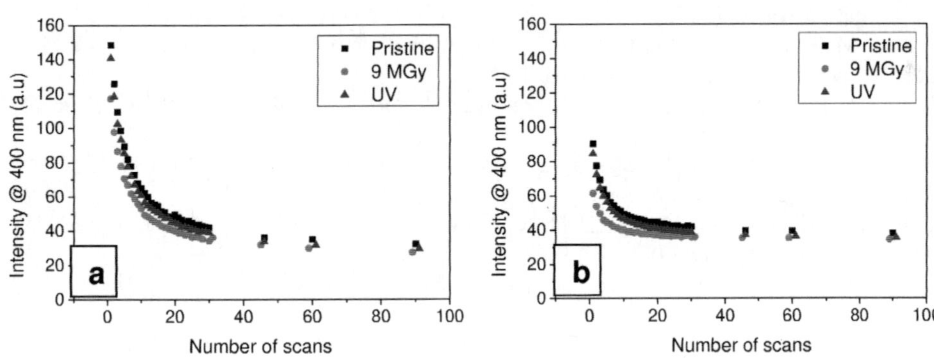

Figure 5. in-situ bleaching kinetics of the GLPC signal under electronic radiation. a. at 3µm away from the center of the core (zone doped with 9 wt% of Ge). b. at 25 µm away from the center of the core (zone doped with 5 wt% of Ge).

4. Conclusion

We reported an in-situ CL investigation of two irradiated samples of germanosilicate OF and we compared their behaviors with a pristine one. Our data indicate that the luminescence observed in all samples is dominated by the emission band at 400 nm attributed to the GLPC point defects. The pre-irradiation with γ-rays up to 9 MGy induced a bleaching of the GLPC intensity of about 30%. Moreover, the GLPC kinetics of the different samples showed that after a sufficiently long electron irradiation, the GLPC concentrations reach a saturation level which is independent from the pre-irradiation. This means that an equilibrium is reached between the conversion and the back conversion processes involving GLPC and the other radiation induced Ge-related defects. Furthermore, we observed that after being exposed to electronic radiation for 30 mins, the GLPC concentration are lower in the Ge-doped region of the core with the higher Ge concentration.

References

[1] Girard S, Kuhnhenn J, Gusarov A, Brichard B, Van Uffelen M, Ouerdane Y, Boukenter A and Marcandella C 2013 *IEEE Trans. Nucl. Sci.* **60** 2015.
[2] Fleming J W 1984 *Appl. Opt.* **23** 4486.
[3] *Defects in SiO2 and Related Dielectrics: Science and Technology*, edited by G. Pacchioni, L. Skuja, and D. L. Griscom (Kluwer Academic, Dordrecht, 2000).
[4] Fujimaki M, Katoh T, Kasahara T, Miyazaki N and Ohki Y 1999 *J. Phys: Condens. Matter* **11** 2589.
[5] Essid M, Albert J, Brebner J. L and Awazu K 1999 *J. Non-Cryst. Solids* **246** 39.
[6] Yamaguchi M, Saito K and Ikushima A. J 2002 *Phys. Rev. B* **66** 132106.
[7] Alessi A, Agnello S, Buscarino G, Pan G, Mashkovtsev R.I, *EPR on Radiation-Induced Defects in SiO2 in Applications of EPR in Radiation research*, Editors A. Lund and M. Shiotani, Springer pp. 255 (2014) 295.
[8] Alessi A, Girard S, Reghioua I, Fanetti M, Di Francesca D, Agnello S, Cannas M, Marcandella C, Martin-Samos L, Richard N, Boukenter A and Ouerdane Y 2015 J. Appl. Phys. **118** 085901.
[9] Di Francesca D, Agnello S, Girard S, Alessi A, Marcandella C, Paillet P, Boukenter A, Gelardi F. M and Ouerdane Y 2016 J. Lightw. Technol **34** 2311.
[10] Skuja L. N 1992 J. Non-Cryst. Solids **149** 77.
[11] Griscom D. L 2011 Opt. Mater. Express **1** 400.
[12] Fujimaki M, Kasahara T, Shimoto S, Miyazaki N, Tokuhiro S, Seol K. S and Ohki Y 1999 *Phys. Rev. B* **60** 4682.
[13] Giacomazzi L, Martin-Samos L, Boukenter A, Ouerdane Y, Girard S and Richard N 2015 *Opt. Mater. Express* **5** 1054.
[14] Yoshikawa M, Matsuda K, Yamaguchi Y, Matsunobe T, Nagasawa Y, Fujino H and Yamane T 2002 J. Appl. Phys **92** 7153.
[15] Goldberg M, Trukhin A and Fitting 1996 H. J Materials Science and Engineering. **B42** 293.
[16] Stevens-Kalceff M. A 2013 Miner Petrol **10** 7455.
[17] Atkins G. R, Poole S. B, Sceats M. G, Simmons H. W and Nockolds C. E 1991 Electronics Letters **27** 1432.
[18] Simmons H. W, Nockolds C. E, Atkins G. R, Poole S. B and Sceats M. G 1992 Micron and Microscopica acta **23** 303.
[19] Atkins G. R, Wang Z. H, McKenzie D. R, Sceats M. G, Poole S. B and Simmons H. W 1993 Lightwave Technology **11** 1793.
[20] Reghioua I, Girard S, Alessi A, Di Francesca D, Martin-Samos L, Fanetti M, Richard N, Raine M, Valant M, Boukenter A and Ouerdane Y 2016 J. Lumin **179** 1.
[21] Medjahdi K, Goutaland F, Boukenter A and Ouerdane Y 2005 J. Non-Cryst. Solids **351** 1835.
[22] Skuja L and Entzia W 1986 phys. stat. sol. (a) **96** 191.
[23] Skuja L, Kajihara K, Hirano M and Hosono H 2011 Phys Rev. B **84** 205206.

IOP Conf. Series: Materials Science and Engineering **169** (2017) 012007 doi:10.1088/1757-899X/169/1/012007

Radiation-induced defects, energy storage and release in nitrogen solids

E Savchenko[1], I Khyzhniy[1], S Uyutnov[1], M Bludov[1], A Barabashov[1] G Gumenchuk[2] and V Bondybey[2]

[1]Institute for Low Temperature Physics & Engineering, NASU, Kharkiv 61103, Ukraine
[2]Lehrstuhl für Physikalische Chemie II TUM, Garching b. München 85747, Germany

elena.savchenko@gmail.com

Abstract. New trends in the study of radiation effects in nitrogen solids with a focus on the defect-induced processes are presented. An electron beam of subthreshold energy was used to generate radiation defects via electronic subsystem. Experimental techniques developed enabled us to detect neutral and charged defects of both signs. Defect production and desorption were monitored using optical and current emission spectroscopy: cathodoluminescence CL, thermally stimulated luminescence TSL and exoelectron emission TSEE along with the detection of postdesorption. Our results show stabilization and accumulation of radiation defects – ionic centres of both signs (N_4^+, N_3^+, N_3^-), trapped electrons and radicals (N, N_3). The neutralization reactions: $N_4^+ + e^- \rightarrow N_4^* \rightarrow N_2^*(a'^1\Sigma_u^-) + N_2^*(a'^1\Sigma_u^-) + \Delta E_1 \rightarrow N_2 + N_2 + 2h\nu + \Delta E_2$ and $N_3^+ + e^- \rightarrow N^*(^2D) + N_2(^1\Sigma_g^+) + \Delta E_3 \rightarrow N(^4S) + N_2(^1\Sigma_g^+) + h\gamma + \Delta E_3$ are shown to be the basis of defect production and anomalous low-temperature post-desorption ALTpD. The part played by pre-existing and radiation-induced defects in energy storage is discussed.

1. Introduction

Nitrogen solids represent model molecular insulating materials with a wide energy gap (15.6 eV). The intense current interest to nitrogen solids is connected with the potential use of polynitrogen compounds as environment-friendly high energy density materials HEDM [1-6]. In ref. [7] it was suggested to synthesize cluster N_4 via excited states of nitrogen. Second important aspect of interest is related to astrophysical research because nitrogen is one of the most abundant elements in the Universe [8]. Solid N_2 is used as moderators [9], source for nitrogen plasma [10] and matrix in radiation chemistry [11]. All these fields of science and applications require an understanding of radiation-induced processes. Despite a long history of solid N_2 spectroscopy such problems as defect production and defect-induced processes in nitrogen solids remained almost unexplored until recently with the exception of N radical formation.

Interaction of ionizing radiation with nitrogen solids results in generation of electron-hole pairs. Electrons are highly mobile in the low-temperature α-phase [12] and their trapping by defect sites or species with positive electron affinity produces negatively charged centers. First observation of trapped electrons in solid nitrogen was reported in [13]. Recently N_3^- centers were detected in solid nitrogen using photon-stimulated exoelectron emission PSEE [14]. Holes in solid N_2 in contrast to electrons have low mobility and can be self-trapped with tetranitrogen cation N_4^+ formation as shown in [15] or trapped on some impurity centers. Creation of ionic species N_3^+ in electron-bombarded nitrogen films was observed by IR and UV absorption [16]. Furthermore besides of N radicals another neutral center – the azid radical N_3, was registered in solid nitrogen pre-irradiated by electrons [17], protons [18] and synchrotron radiation [19]. Indication of a number of radiation-induced defect centers poses the question on their stability, dynamics and relaxation processes accompanied by energy transfer and conversion.

In view of such a diversity of the centers, which can be generated by irradiation, we applied a complex of complementary techniques to be able to explore the entire set of defects, their relaxation and related reactions with a focus on particles ejection from pre-irradiated solid N_2. Cathodoluminescence CL spectroscopy as well as optical and current activation spectroscopy methods: thermally stimulated luminescence TSL and exoelectron emission TSEE along with the detection of the total yield of post-desorption, which occurs upon heating, have been applied. This enabled us to get information on radiation-induced defects and elucidate their role in defect-induced processes taking place in an electron-bombarded solid N_2.

2. Experimental technique

We have created a comprehensive technique that enabled us to detect both neutral and charged defects of both signs. The developed experimental technique has previously been described [20, 21] therefore only important details related to the present study are given below.

2.1. Sample preparation and generation of defects

The nitrogen solids were grown by deposition of high purity (99.995%) N_2 gas onto a cooled Cu substrate mounted in a high-vacuum chamber with a base pressure of 10^{-8} mbar. A convenience of the sample preparation from the gas phase is the ability to obtain samples of the required thickness with an open surface. Such an open surface makes it possible to monitor particles desorption and defect production with luminescence techniques in a wide range of wavelength from infrared IR to vacuum ultraviolet VUV and apply current activation techniques based on exoelectron emission – TSEE and PSEE.

To irradiate the samples we used a low-energy electron beam to ensure the generation of defects through electronic subsystem, so-called sub-threshold mode. Taking into account that the threshold electron beam energy E_{thr} for production of defects in solid N_2 via the knock-on mechanism is about 1.5 keV we used beams with energies in the range 300-1500 eV. The irradiation was performed in dc regime. The current density was 3 mAcm^{-2}. The sample temperature was controlled with a Si sensor. The sample heating under electron beam did not exceed 0.5 K. The CL spectra were detected concurrently in the visible range and in VUV with two spectrometers. To measure the dose dependence of excited species production the CL spectra were recorded repeatedly on an exposure time or the CL intensity was recorded at the selected wavelength as a function of exposure time. Note that luminescence techniques provide access to forbidden electronic states which are difficult to populate using conventional absorption spectroscopy.

2.2 Measurement of relaxation emissions: photons, electrons and particles

Specific feature of nitrogen solids is the existence of a very long-lived metastable state of the nitrogen atoms, which appear under irradiation, resulting in long afterglow which was observed on completion of the irradiation at the wavelength of the forbidden atomic transition $^2D \rightarrow {}^4S$ [22]. Simultaneously with the afterglow we observed an "afteremission" of electrons with similar characteristic decay times.

When the afterglow and "afteremission" current had decayed to essentially zero, three relaxation emissions were measured in a correlated fashion – emission of photons, electrons and particles. Relaxation processes were studied at stimulation by heating. In these experiments we used heating with a constant rate, for the most part 5 Kmin^{-1}. Partial yields of TSL were measured in the visible and VUV ranges. The measurements of thermally stimulated relaxation emissions were performed in the temperature region 5–30 K. Being promoted to the conduction band by heating detrapped electrons either neutralize positively charged centers yielding TSL or escape from the sample yielding TSEE. Stimulated currents were detected with an electrode kept at a small positive potential V_F=+9 V and connected to the current amplifier. Coincidentally with TSEE and TSL detection the particles ejection (post-desorption) was recorded by pressure P monitoring. The entire control of the experiment was made with the help of a computer program developed specifically for these studies.

3. Results and discussion

Luminescence spectra in conjunction with activation spectroscopy provide extensive information on radiation-induced transformation of solids under ionizing radiation: molecule fragmentation or association, accumulation of radiation–induced species, modification of structure, desorption of excited particles, radiation-induced chemical reactions and cascades of relaxation processes involving both electronic and atomic processes. Figure 1 demonstrates the CL spectrum of solid N_2 measured in the VUV range. It consists of two forbidden molecular progressions the $a'^1\Sigma_u^- \rightarrow X^1\Sigma_g^+$ and $A^3\Sigma_u^+ \rightarrow X^1\Sigma_g^+$.

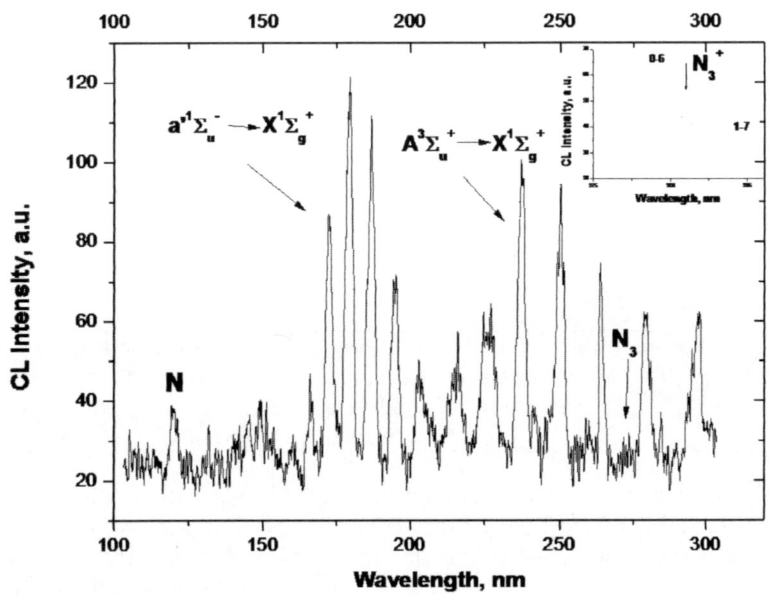

Figure 1. CL spectra of solid N_2 recorded at 5 K. Inset shows a part of spectrum recorded with higher resolution. Features related to N_3 and N_3^+ species are marked by arrows.

The vibrational bands positions of both progressions, are shifted toward longer wavelength with respect to the gas phase spectra, and are in good agreement with those observed in the early studies of VUV luminescence of solid N_2 [23, 24]. As it was found in [15] both progressions show dose dependence on an exposure time indicating accumulation of radiation-induced defects responsible for the emissions observed, which are thought to be of ionic nature. Another indication of the ionic nature of primary "reagent" in the neutralization reaction with electron was the observation TSL, nonstationary luminescence NsL and photon/stimulated luminescence PSL on vibronic bands of the singlet transition $a'^1\Sigma_u^- \rightarrow X^1\Sigma_g^+$, the yields of which correlated with exoelectron current emissions TSEE and PSEE [14, 15]. Behaviour of the relaxation emissions together with analysis of the energy diagram and potential curves of N_4 available [1, 7] resulted in the conclusion that the neutralization of tetranitrogen cation N_4^+, in other words self-trapped hole, causes its dissociation followed by the exit

IOP Conf. Series: Materials Science and Engineering **169** (2017) 012007 doi:10.1088/1757-899X/169/1/012007

(from the parent "cage" where the neutralization has occurred) of molecules in the lowest excited singlet state $a'^1\Sigma_u^-$, which then decays radiatively:

$$N_4^+ + e^- \rightarrow N_4^* \rightarrow N_2^*(a'^1\Sigma_u^-) + N_2^*(a'^1\Sigma_u^-) + \Delta E_1 \rightarrow N_2 + N_2 + 2h\nu + \Delta E_2 \quad (1)$$

The VUV CL spectrum has several features first detected by the CL technique. This is a weak band at 273 nm assigned to the $A^2\Sigma_u^+(000) \rightarrow X^2\Pi_g(000)$ transition of N_3 radical in accordance with [16] where this transition was observed in VUV absorption spectrum of solid N_2 exposed to energetic electrons. Very low intensity of this band in our experiments is likely due to efficient formation of N_3^- centers detected with PSEE technique in our recent experiments [14]. A high value of N_3 electron affinity E_e (2.76 eV) promotes the electron attachment reaction. More intense feature which was observed in the VUV CL spectrum of higher resolution (shown in the inset in Figure 1) is the band at 281 nm. This band belongs to the transition $A^3\Pi_u(000) \rightarrow X^3\Sigma_g^-(000)$ of N_3^+ cation and coincides with the band detected in the VUV absorption of electron bombarded N_2 [16]. We detected also the atomic line N in the VUV CL spectrum which was identified as $3s\ ^4P \rightarrow 2p^3\ ^4S$ transition emitted by desorbing N atoms (this issue will be addressed in a subsequent publication).

Detection of the total yield of particles from pre-irradiated solid N_2 (first observed in [25]) clearly

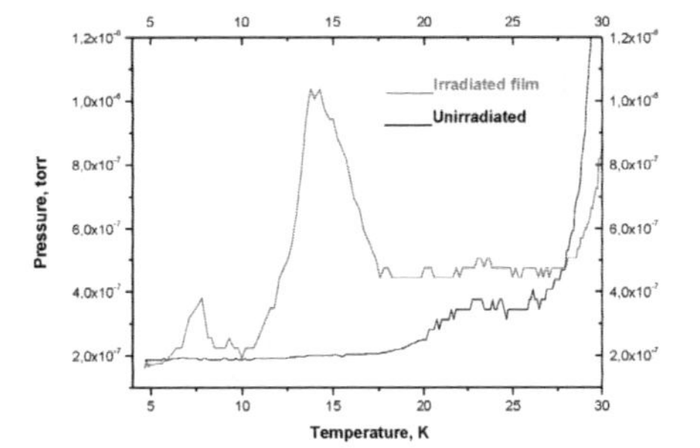

Figure 2. P behaviour upon warm-up of the unirradiated film of solid N_2 (low curve) and the film pre-irradiated with an electron beam (upper curve).

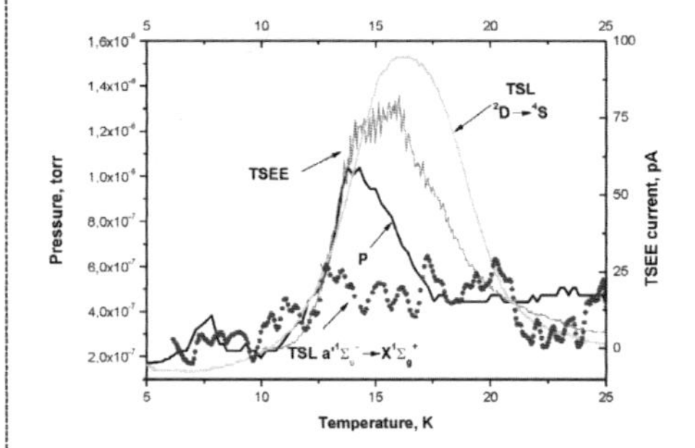

Figure 3. Temperature dependence of P detected synchronously with the TSEE and TSL registered at wavelengths of the transitions $^2D \rightarrow ^3S$ and $a'^1\Sigma_u^-(0) \rightarrow X^1\Sigma_g^+(5)$.

demonstrates the radiation nature of this phenomenon. Comparison of the particle yields by P measurements from unirradiated and pre-irradiated solid N_2 films of 100 μm thick is shown in Figure 2. Note that both samples were not annealed before the experiment. For the unirradiated sample there is nearly no change in pressure up to 20 K and the sample starts to sublimate at about 28 K. A sharp rise in pressure followed its decrease upon warm-up of the pre-irradiated sample was observed in the region of temperatures much lower than the characteristic sublimation temperature of solid N_2. The most likely source of energy needed for post-desorption may be neutralization of positively charged centers survived after completion of irradiation, which are N_4^+ and N_3^+ centers. The relaxation channel described by the reaction (1) is followed by an energy release, at the stage after neutralization as well as after the radiative transition $a'^1\Sigma_u^- \rightarrow X^1\Sigma_g^+$. Note that the transition to the ground state (shown in Figure 1) terminates in vibrationally excited levels with a maximum of vibrationally excited molecules distribution at 5th level, that corresponds to about 1.45 eV. Neutralization of N_3^+ center registered in the CL spectrum also proceeds with a considerable energy release. As it was found in [26] the most probable channel of the N_3^+ dissociative recombination is the two-body channel. If both products are in the ground state the energy release in this reaction exceeds 10 eV [26], which creates preconditions for dissociation with the creation of $N(^2D)$ center. We observed accumulation of these centers upon irradiation [27] by the reaction:

$$N_3^+ + e^- \rightarrow N^*(^2D) + N_2(^1\Sigma_g^+) + \Delta E_3 \rightarrow N(^4S) + N_2(^1\Sigma_g^+) + h\gamma + \Delta E_3 \quad (2)$$

In order to validate the suggestion on connection of ALTpD with neutralization reactions we measured simultaneously P, TSEE and TSL on the wavelengths of the transitions connected with reactions (1) and (2) with the result shown in Figure 3. Detection of the TSEE yield indicates electron detrapping. Then recombination of electrons with positively charged nitrogen centers occurs followed by an energy release. The same threshold for ALTpD yield (P) and yields of TSEE and TSL, detected on the wavelengths of the transitions involved in the reaction (1) and (2) indicates that ALTpD is caused by the neutralization reactions.

To determine which traps are sources of electrons to recombine with N_4^+ and N_3^+ centers we carried out an experiment with irradiation cycles. Annealed film of solid N_2 10 μm thick was subjected to the cycles: (i) irradiation by 1.4 keV electrons, then (ii) warm-up to 25 K (while monitoring the pressure) and re-cooling to 5 K. In each subsequent cycle the sample was irradiated for a longer time and curves taken after different irradiation times are shown in Figure 4. At low irradiation time pre-exising traps (growth defects) are filled and two peaks at about 16 and 11 K are detected.

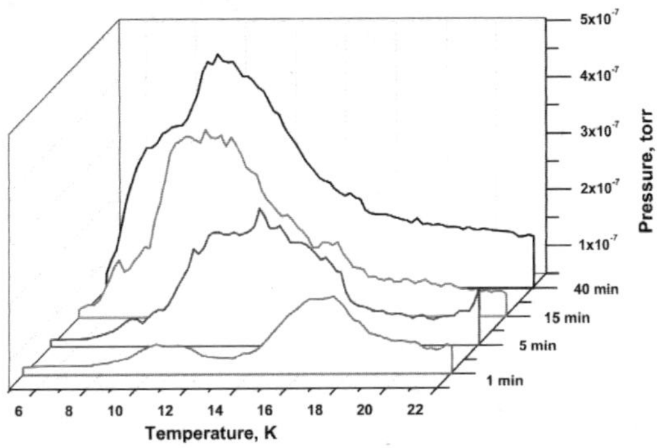

Figure 4. ALTpD yields (P) measured after irradiation during different times.

With increasing irradiation time other traps associated with radiation-induced defect centers are populated. Nonelementary curve of pressure indicates a set of defects involved. Taking into account that in a subthreshold regime defects can be formed only via electronic subsytem these radiation induced defects are most likely related to the lattice distorsion at formation of atomic centers N, and centers, containing more than two N atoms. The radiation-induced traps appeared to be more shallow than that associated with pre-existing defects. At present it is difficult to identify separate peaks in the curves taken at irradiation times longer than 1 min (Figure 4), the trap levels distribution looks a nearly continuous. N_3^- centers representing deap traps do not contribute to ALTpD. The information obtained is of importance for different cryogenic systems in radiation environment, e.g. for astrophysical objects, moderators and for operation conditions of big particle accelerators like Large Hadron Collider at CERN where cryogenic surfaces are subjected to intense ionizing radiation.

4. Summary

The set of radiation-induced defects in N_2 solids pre-irradiated with an electron beam was detected and studied with the focus on the role of charge centers in the radiation-induced processes. The CL detection and correlated in real time measurements of spectrally resolved thermally stimulated luminescence TSL and exoelectron emission TSEE as well as detection of thermally stimulated particle ejection from pre-irradiated N_2 films were performed. This enabled us to identify N_4^+, N_3^+, N_3^-, N_3 and N radiation-induced defect centers in nitrogen solids exposed to an electron beam. Connection of strong anomalous low-temperature post-desorption ALTpD from pre-irradiated solid N_2 with radiation-induced defects was proved and underlying this phenomenon mechanism – the dissociative recombination of charged centers N_4^+ and N_3^+ with detrapped electrons, was elucidated.

Acknowledgments

The authors cordially thank Profs. Peter Feulner, Minh Tho Nguyen, Aleksands Lushchik and Giovanni Strazzulla for stimulating discussions.

5. References

[1] Nguyen M T 2003 *Coord. Chem. Rev.* **244**
[2] Eremets M I, Trojan I A, Gavriliuk A G and Medvedev S A 2008 *Synthesis of High-Nitrogen Energetic Material* in *Static Compression of Energetic Materials* ed Peiris S M and Piermarini G J (Berlin Heidelberg: Springer-Verlag) p 75
[3] Zahariev F, Hooper J, Zhang F, Woo T K 2007 *Phys. Rev. B* **75** 140101
[4] Ma Y, Oganov A R, Li Z, Xie Yu, and Kotakoski J 2009 *Phys. Rev. Lett.* **102** 065501
[5] Zarko V E 2010 *Combustion, Explosion, and Shock Waves* **46** 121
[6] Yakub L N 2013 Low Temp. Phys. **39** 427
[7] Elesin V F, Degtyarenko N N, Pazhitnykh K S, Matveev N V 2009 *Russ. Phys. J.* **52** 1224
[8] Clark R N, Carlson R, Grundy W, Noll K N 2012 *Observed Ices in the Solar System*, in *The Science of Solar System Ices* ed Gudipati M S and Castillo-Rogez J, Astrophysics and Space Science Library, Vol. 356 (New York: Springer) p 3
[9] Ghandi K and Miyake Y 2011 *Muon interaction with matter*, in: *Charge Particle and Photon Interaction with Matter: Recent advances, applications and interfaces* ed Hatano Y Katsumura Y and Mozumder A (Boca Raton, London, New York: CRC Press) p 169
[10] Niino H, Sato T, Narazaki A, Kawaguchi Y, Yabe A 2002 *Appl. Surf. Sci.* **197-198** 67
[11] Grigoriev E I and Trakhtenberg L I 1996. *Radiation-Chemical Processes in Solid Phase: Theory and Application.* (Boca Raton: CRC Press) p 240
[12] Storchak V G, Eshchenko D G, Brewer J H, Cottrell S P, Cox S F J, Karlsson E, Wappling R W, 2001 *J. Low Temp. Phys.* **122** 527
[13] Khyzhniy I V, Savchenko E V, Uyutnov S A, Gumenchuk G B, Ponomaryov A N, Bondybey V E, 2010 *Radiation Measurements* **45** 353
[14] Savchenko E, Khyzhniy I, Uyutnov S, Bludov M, Gumenchuk G and Bondybey V 2016 *Radiation Measurements* **90** C 1
[15] Savchenko E V, Khyzhniy I V, Uyutnov S A, Barabashov A P, Gumenchuk G B, Beyer M K, Ponomaryov A N and Bondybey V E 2015 *J. Phys. Chem.* A **119**, 2475
[16] Wu Y-J, Chen H-F, Chuang S-J, Huang T-P 2013 *Astrophys. J.* **768** 83
[17] Jamieson C S and Kaiser R I 2007 *Chem. Phys. Lett.* **440** 98
[18] Hudson R L and Moore M H 2002 *Astrophysical Journal* **568** 1095
[19] Wu Y-J, Wu C Y R, Chou S-L, Lin M-Y, Lu H-C, Lo J-I, Cheng B-M 2012 *Astrophys. J.* **746** 175
[20] Savchenko E V and Dmitriev Yu A 2010. *New Aspects of Relaxation Processes in Cryogenic Solids,* in: *Applied Physics in the 21st Century* ed Raymond P. Valencia (Horizons in World Physics. Volume 269) (New York: Nova Science Publishers) p 113
[21] Allodi M A, Baragiola R A, Baratta G A, Barucci M A, Blake G A, Brucato J R, Contreras C, Cuylle S H, Boduch Ph, Fulvio D, Gudipati M S, Ioppolo S, Kaňuchová Z, Lignell A, Linnartz H, Palumbo M E, Raut U, Rothard H, Salama F, Savchenko E V, Sciamma-O'Brien E, Strazzulla G, ed Strazzulla G, 2013 *Space Sci. Rev.* **180** 101
[22] Lofthus A and Krupenie P H 1977 *J. Phys. Chem. Ref. Data* **6** 113
[23] Fugol' I Ya, Poltoratskii Yu B, Savchenko E V 1976 *JETP Lett.* **24** 1
[24] Coletti F and Bonnot A M 1977 *Chem. Phys. Lett.* **45** 580
[25] Savchenko E V, Khyzhniy I V, Uyutnov S A , Ponomaryov A N, Gumenchuk G B, Bondybey V E 2013 *Low Temp. Phys.* **39** 574
[26] Zhaunerchyk V, Geppert W D, Vigren E, Hamberg M, Danielsson M, Larsson M, Thomas R D, Kaminska M, Österdahl F 2007 J. *Chem. Phys.* **127** 014305
[27] Savchenko E, Khyzhniy I, Uyutnov S, Barabashov A, Gumenchuk G, Ponomaryov A and Bondybey V 2015 *Phys. Status Solidi* (c) **12** 49

Irradiation temperature effects on the induced point defects in Ge-doped optical fibers.

A.Alessi[1], I. Reghioua[1], S.Girard[1], S. Agnello[2], D. Di Francesca[1], L.Martin-Samos[3], C. Marcandella[4], N.Richard[4], M. Cannas[2], A.Boukenter[1], Y. Ouerdane[1]

[1] Univ-Lyon, Laboratoire H. Curien, UMR CNRS 5516, Université Jean Monnet, 18 rue du Pr. Benoît Lauras, 42000 Saint-Etienne, France.
[2] Dipartimento di Fisica e Chimica, Università di Palermo, I-90123 Palermo, Italy.
[3] CEA, DAM, DIF, F91297, Arpajon, France.
[4] Materials Research Laboratory, University of Nova Gorica, Vipavska 11c 5270-Ajdovscina, Slovenija.

antonino.alessi@univ-st-etienne.fr

Abstract. We present an experimental investigation on the combined effects of temperature and irradiation on Ge-doped optical fibers. Our samples were X-ray (10 keV) irradiated up to 5 kGy with a dose rate of 50 Gy(SiO_2)/s changing the irradiation temperature in the range 233-573 K. After irradiation we performed electron paramagnetic resonance (EPR) and confocal microscopy luminescence (CML) measurements. The recorded data prove the generation of different Ge related paramagnetic point defects and of a red emission, different from that of the Ge/Si Non-Bridging Oxygen Hole center. Furthermore, by comparing the behaviour of the EPR signal of the Ge(1) as a function of the irradiation temperature with the one of the red emission we can exclude that this emission is originated by the Ge(1).

1. Introduction

The understanding of point defects properties and generation in pure and doped silica has been at the center of many experimental and simulative investigations for many years [1-5]. On one side, the study of optical and magnetic properties of point defects in amorphous materials represents a significant field within solid state physics, on the other side, the use of silica as basic material in different application areas as nanostructured systems [6] and optical fibers [7], just to cite some of them, has imposed the understanding of the impact of such defects on the physical properties of silica and as a consequence on silica based devices. Clear examples of this fact are the alterations of the transmission properties of the optical fibers as a consequence of the generation of point defect species having optical absorption bands [7] or the increase of the silica sensitivity to radiation due to absorbing defects related to doping or manufacturing processes [8,9].

Ge related defects in silica have been widely investigated, indeed, Ge atoms can be easily found as impurity in natural silica [10] and the Ge doping is commonly employed to increase the silica refractive index [11] and as a consequence it is used for the production of optical fibers and fiber based devices [7]. Nowadays it is know that the Ge atoms can form different point defects, one of them is the so called GLPC (Germanium Lone Pare Center [2]) involved in many radiation induced processes [2,8,9], which are the origin of the photosensitivity of this silica typology. Other known defects are the E'Ge, the Ge(1) and the Ge(2) [1-3]. For these defects various investigations have been performed to elucidate the generation or conversion processes under irradiation at room temperature [2,4,5,8,12] or to identify their thermal stability with post irradiation treatments [13]. On the other side combined temperature and irradiation effects have been poorly studied [14]. Similarly, an emission activity induced in the red spectral domain received low attention. Basing on published data, in this spectral range it is found the luminescence of the Ge/Si Non Bridging Oxygen Hole Center [15], anyway, this luminescence is not isolated. Indeed, the authors of [16, 17] observed in H_2 loaded silica samples another red emission attributing this activity to the H(II) (a GLPC forming a linkage with an hydrogen atom [10,16] through one of the two electrons of the lone pair). More recently [18], we observed a red emission in dissimilar Ge-doped silica. Such luminescence presents some differences with respect to that reported in [16, 17] and it was observed in bulk and fiber samples in which the H(II) EPR signal was not detected. In that study it was shown that this emission is not related to E'Ge or Ge(2) defects, but relation with the Ge(1) center was not rigorously ruled out. In the present study we report on the effects of temperature and irradiation on the generation of Ge-related point defects in Ge-doped fibers. This thermal and irradiation concurrent study enabled to find out the absence of correlation between Ge(1) and red emission paving the way to the attribution of this latter activity to other defects.

2. Experimental

We studied a two steps multimode Ge-doped optical fiber produced by iXBlue Photonics Division. The starting preform was produced by MCVD (Modified Chemical Vapor Deposition) process.

As previously reported [19] the sample has a total diameter of 125 μm with a Ge doped part of ~60 μm. More in details, in the central core, having diameter of ~40 μm, the Ge doping level is ~8 weight per cent (% wt), whereas the second step, being the external ring of the doped part, has a doping level of ~4.5 % wt.

The ARACOR and PROBIX machines located at the French atomic energy center (CEA) were employed to perform X-ray (10 keV) irradiations up to 5 kGy with a dose rate of 50 Gy(SiO_2)/s. The irradiation temperature was imposed by keeping the fibers in contact with a temperature controlled plate.

A Bruker EMX-Micro Bay spectrometer operating at ~9.8 GHz (X-band) with a 100 kHz modulating magnetic field was employed to record EPR measurements. Microwave power and modulation amplitude were opportunely selected to avoid signal distortion.

CML spectra were recorded using the 442 nm laser line of an Aramis (Jobin-Yvon) spectrometer equipped with a micro-translational stage and with a CCD camera. All the data were recorded in a back-reflection geometry using a 50× objective, with a spatial resolution of ~3 μm.

3. Result and Discussion

In figure 1 we report the double integral (normalized to the sample masses and to the instrumental parameters) of the EPR signal acquired in the different samples as a function of the irradiation temperature for the dose 5kGy. The data indicate that the total amount of induced defects decreases with increasing the irradiation temperature. Furthermore, as illustrated by the two spectra, reported as insets in figure 1 and acquired for the samples irradiated at 233 and 570 K, the lineshape of the spectra is not constant.

More in details, the spectra of all the samples result from the combination of the signals of the Ge(1), Ge(2) and E'Ge defects [20] but the relative contributions of the different species depend on the irradiation temperature, in particular we note that increasing the irradiation temperature the signal lineshape tends to the one of the E'Ge. Such data together with the decrease of the EPR signal proves the decrease of the amount of Ge(1) defects as a function of the irradiation temperature. Finally, we note that we did not detect the characteristic EPR signal of the H(II) defects [16,17].

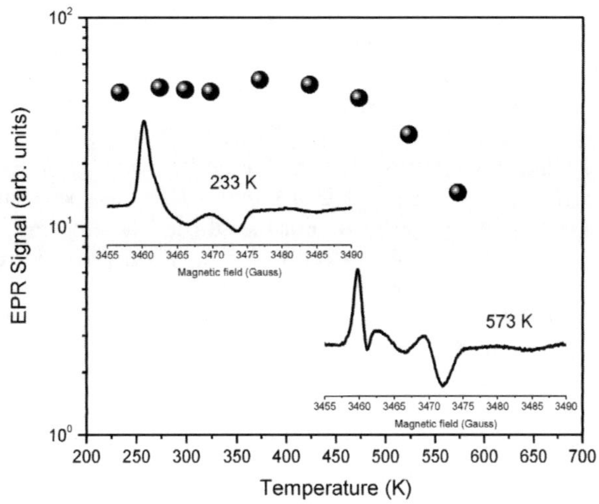

Figure 1. Double integral of the EPR signal as a function of the irradiation temperature at the dose 5 kGy. Insets EPR spectra recorded for the samples irradiated at 233 and 573 K.

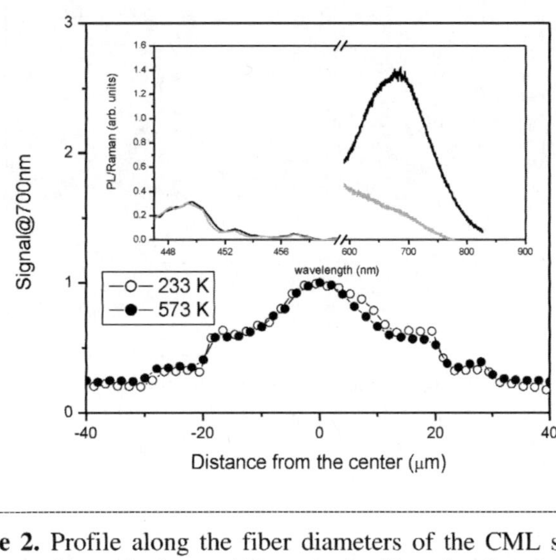

Figure 2. Profile along the fiber diameters of the CML signal at 700 nm under excitation at 442 nm in the fibers irradiated at 233 K (-○-) and 523 K (-●-). Inset PL/Raman signal in the core of the pristine (—) and of the 573 K irradiated fiber (—). Dose 5 kGy.

In figure 2, we report the radial profile of the photoluminescence (PL) signal measured at 700 nm under excitation at 442 nm for the samples irradiated at the lowest and highest temperature. The signal features a maximum at the center of the core. This maximum is related to the presence of an emission band peaked at about 680 nm. In the inset of the same figure we illustrate the PL signals of the pristine and 573 K irradiated samples. All the spectra were normalized to the silica Raman signal [21] located in the range 447-460 nm. We note that these data agree with previous investigations [18].

Figure 3 illustrates the spectra recorded for samples irradiated at different temperatures. Also in this case all the measurements were normalized to the silica Raman signals located at low wavelengths. The data indicate that the relative amplitude (with respect to the Raman signal) of the emission band at 680 nm is independent from the irradiation temperature up to about 420 K. Then, its amplitude increases, we also note that other signal at low wavelengths increases up to 523 K and then it decreases at the highest irradiation temperature. For the aims of the present investigation we focused our attention on the behaviour of the 680 nm component and on possible relation with the Ge(1) defects. In the inset of figure 3, we report the estimation of the relative amplitude at 680 nm as a function of the irradiation temperature after the subtraction of a line tangent to the spectra at about 590 and 825 nm. These data confirm that the amplitude of this band starts to increase for irradiation temperature higher than 420 K. We note that this temperature is very close to the one at which the EPR signal starts to decrease (see fig.1) and that the behaviour of its amplitude is the opposite deduced for the induced concentration of the Ge(1). As a consequence we can conclude that the Ge(1) is not responsible for the 680 nm emission band. On the other side, further investigation should be performed to better clarify the fact that both signals start to change at similar temperature and try to connect this effect to microscopic mechanisms.

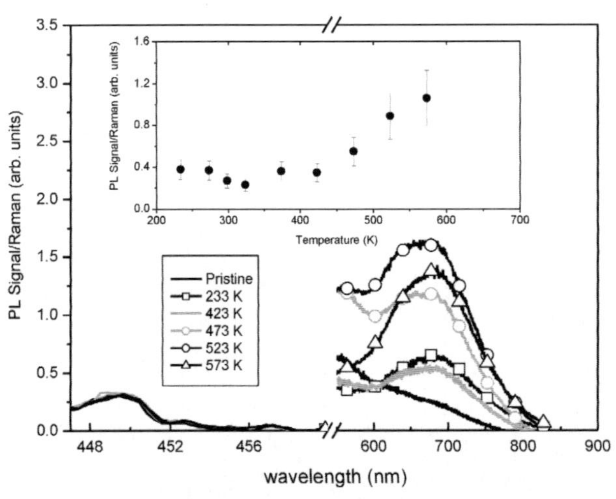

Figure 3. PL/Raman signals recorded in the center of core of pristine sample (—) and of fibers irradiated at 233 K (-□-), 423 K (—), 473 K (-○-), 523 K (-○-) and 573 K (-Δ-). Dose 5 kGy.

4. Conclusion

We studied the radiation effects on a multimode Ge doped fiber irradiated at different temperatures. In particular, we focused our attention on the behavior of the induced Ge related paramagnetic defects and of a red luminescence activity. Our data indicate that the total amount of the former decreases with increasing the irradiation temperature, whereas the amplitude of the emission increases. Furthermore, since the modification of the line-shape of the electron paramagnetic resonance spectra clearly indicates the decrease of the induced Ge(1) we can exclude that this type of defects are the origin of the red emission peaked at 680 nm under excitation at 442 nm.

References

[1] Friebele E J and Griscom D L 1986 *Mat. Res. Soc. Symp. Proc.* **61** 319.
[2] Fujimaki M, Kasahara T, Shimoto S, Miyazaki N, Tokuhiro S, Seol K S and Ohki Y 1999 *Phys. Rev. B* **60** 4682.
[3] Giacomazzi L, Martin-Samos L, Boukenter A, Ouerdane Y, Girard S and Richard N 2015 *Opt. Mater. Express* **5** 1054.
[4] Chiodini N, Meinardi F, Morazzoni F, Paleari A and Scotti R 1999 *Phys. Rev. B* **60** 2429.
[5] Cuevas R F, Sekiya E H, Garcia-Quiroz A, Da Silva E C and Suzuki C K 2006 *Nucl. Instrum. Method B* **247** 285.
[6] Alessi A, Agnello S, Buscarino G, Cannas M, Gelardi F M, Sporea A, Sporea D and Vâţă I 2014 *J. Mater. Sci.* **49** 6475.
[7] Girard S, Kuhnhenn J, Gusarov A, Brichard B, Van Uffelen M, Ouerdane Y, Boukenter A and Marcandella C 2013 *IEEE Trans. Nucl. Sci.* **60** 2015.
[8] Alessi A, Girard S, Cannas M, Agnello S, Boukenter A and Ouerdane Y 2011 *Opt. Exp.* **19** 11680.
[9] Alessi A, Fanetti M, Agnello S, Girard S, Buscarino G, Di Francesca D, Reghioua I, Messina F, Cannas M, Martin-Samos L, Valant M, Richard N, Boukenter A and Ouerdane Y 2016 *Opt. Mater. Express* **6** 2213.
[10] Defects in SiO_2 and Related Dielectrics: Science and Technology edited by G. Pacchioni, L. Skuja, and D. L. Griscom (Kluwer Academic,Dordrecht, 2000).
[11] Fleming J W 1984 *Appl. Opt.* **23** 4486.
[12] Di Francesca D, Agnello S, Girard S, Alessi A, Marcandella C, Paillet P, Boukenter A, Gelardi F M, , Ouerdane Y 2016 *J. Lightwave Tech.* **34** 2311.
[13] Griscom D L 2011 *Opt. Mater. Express* **1** 400.
[14] Girard S, Marcandella C, Morana A, Perisse J, Di Francesca D, Paillet P, Macé J-R, Boukenter A, Leon M, Gaillardin M, Richard N, Raine M, Agnello S, Cannas M, Ouerdane Y, 2013 *IEEE Trans. Nucl. Sci.* **60** 4305.
[15] Skuja L and Naber A 1996 *Nucl. Instrum. Method B* **116** 549.
[16] Awazu K, Muta K-I and Kawazoe H 1993 *J. Appl. Phys.* **74** 2237.
[17] Poulios D P, Spoonhower J P and Bigelow N P 2003 *J. Lumin.* **101** 23.
[18] Alessi A, Di Francesca D, Agnello S, Girard S, Cannas M, Richard N, Boukenter A and Ouerdane Y, 2016 *J. Lumin.* 177 127.
[19] Alessi A, Girard S, Marcandella C, Agnello S, Cannas M, Boukenter A and Ouerdane Y 2011 *J. Non-Cryst. Solids* **357** 1966.
[20] Alessi A, Agnello S, Gelardi F M, Messina G and Carpanese M 2011 *J. Non-Cryst. Solids* **357** 1900.
[21] Geissberger A E and Galeener F L 1983 *Phys. Rev. B* **28** 3266.

Radiation effects on microstructure and EPR signal of yttrium oxide rods

S C Santos, O Jr Rodrigues , L L Campos

Instituto de Pesquisas Energeticas e Nucleares – IPEN, Av. Prof. Lineu Prestes 2242, Cidade Universitaria, Sao Paulo, Brazil.

silas.cardoso@usp.br

Abstract. Designing nanostructured materials with high dosimetric efficiency is a great challenge in radiation dosimetry research. From rare-earth series, yttrium oxide is considered as excellent host matrix for rare-earth ions, leading to formation of advanced functional materials with optical, mechanical, chemical, and thermal properties notably improved. Nevertheless, there is a lack of information which correlates microstructural characteristics and performance of rare earths. This work aims to evaluate the radiation effects on microstructure and EPR signal of Y_2O_3 rods produced by colloidal processing followed by sintering at 1600°C/4h in air. Ceramic rods were exposed to gamma radiation with doses up to 100kGy. Microstructural and dosimetric characterizations were performed by XRD, SEM and EPR techniques. Yttrium oxide rods as sintered exhibited dense microstructure (96.6% theoretical density) and linear EPR dose response behaviour for wide dose range. These results reveal that yttrium oxide is a promising material for radiation dosimetry.

1. Introduction

Innovation in design of dosimetric materials comprises synthesis, processing and conformation of nanostructured components. In addition, the properties of solid dosimetric materials depend on chemical composition, crystal lattice and microstructure, which is the arrangement of the atoms, phases and defects within a material.

The interaction of ionizing radiation (α, β, γ rays) with a solid can induce physical modifications such as excitation (ionization) of atoms and molecules, a pair of electron- hole that leads either recombination luminescence centres or to formation of an interstitial atom and vacancy. A powerful technique for characterization of ionizing induced defects is Electron Paramagnetic Resonance (EPR). In addition, EPR is used to evaluate polymerization reactions[1], catalysis[2], laser[3], free radicals in living tissues and fluids[4] and drug detection[5]. EPR is a useful technique for solid state dosimetry.

Yttrium oxide (Y_2O_3) belongs to rare earth sesquioxides series and exhibits remarkable properties, which make it a promising material for radiation dosimetry. Further, Y_2O_3 applications include sintering aid[6], catalysis[7], luminescence[8], electrical[9], electronic[10], mechanical[11] and thermal[12] materials. Even though Y_2O_3 presents great applicability, there is a lack of investigations on radiation effects in yttrium oxide based ceramics. The purpose of this paper is to evaluate radiation induced effects on microstructure evolution and EPR response of yttrium oxide rods produced by colloidal processing.

2. Experimental

For this work, yttrium oxide powders (Y_2O_3 99.99%; Johnson Matthey) with mean particle size (d_{50}) of 190nm, specific surface area (SSM) of $6.4m^2.g^{-1}$, pycnometric density (ρ) of $4.85g.cm^{-3}$ were used as raw material. Yttrium oxide rods were formed by colloidal processing using suspensions with 30vol% of Y_2O_3 particles prepared at pH 10 through tetramethylammonium hydroxide solution (TMAH, 25wt.% in water, Sigma Aldrich). All procedures and parameters for stabilization of Y_2O_3 suspensions are reported in our previous study[13]. The homogenization of ceramic suspensions was performed with a ball mill for 24h using alumina spheres ($Ø_{spheres}$=10mm). Ceramic samples (4.35 x 2.27mm, height x diameter) were shaped as micro rods by casting using organic templates on a plaster plate as illustrated in Figure 1. Thermal treatment of conformed samples was realized in a vertical furnace (Lindberg Blue) at temperature of 1600°C for 4h in ambient atmosphere. Crystalline structure of as sintered samples was evaluated by X-ray diffraction (XRD; Rigaku Multiflex), scanning at $1^{\circ}.min^{-1}$, $\Delta\theta = 10\text{-}80^{\circ}(2\theta)$, radiation Cu-kα and identified according to powder diffraction files (PDF).

Batches of four yttrium oxide rods as sintered were irradiated with gamma source with doses from 1Gy to 100kGy in ambient temperature. Microstructural evaluation of irradiated samples was performed by Scanning electron microscopy (SEM, Incact-X Oxford). Crystal defects and radicals induced by ionizing radiation were characterized by electron paramagnetic resonance at ambient temperature and atmosphere using X-band EPR spectrometer (Bruker EMX PLUS). EPR spectra of samples were recorded using the following parameters: field modulation frequency of 100kHz, microwave power of 0.6325mW, center field at 3200G, sweep width of 6000 G, modulation amplitude of 4 G, time constant of 0.01ms and, 10 scans. The EPR spectra of irradiated samples were determined as a mean of each batch normalized by mean mass of samples. EPR dose response and time decay curves were plotted considering the mean of peak-peak amplitudes of irradiated samples.

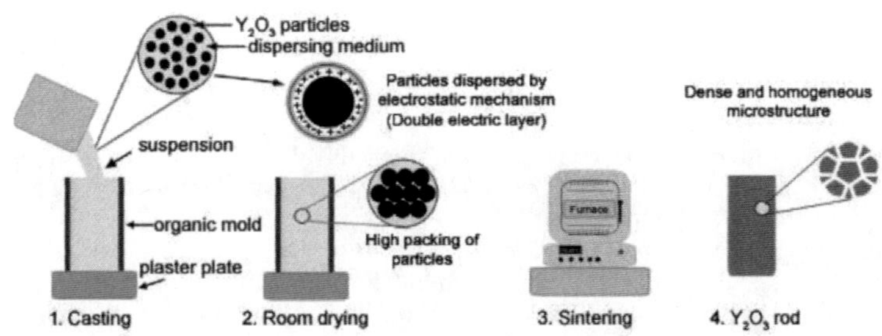

Figure 1 – Sequential colloidal processing for production of yttrium oxide rods.

3. Results and discussion

In ceramic processing there is a great concern on material structure, which has a profound effect on properties as well as on performance of the final product. The first type of structure is at atomic scale i.e. type of bonding and crystal structure (arrangement of atoms). Thus, crystal characterization is primordial step to development advanced materials for radiation dosimetry. Fig.2a shows XRD pattern of Y_2O_3 rods (Fig.2b). A characteristic diffraction peak of high intensity at 29° is observed, which is related with the crystallographic plane (222), corresponding to body-centered cubic yttria (C-type) and fitting the PDF (70-603) standard. Hoekstra et al.[14] reported that rare earths sesquioxides belong to C-type as Dy_2O_3, Th_2O_3, Ga_2O_3 and In_2O_3. In addition, literature has shown that Y_2O_3 can present other polymorphs. Gourlaouen et al.[15] reported the monoclinic structure (B-type) at 997°C under 2.0 GPa during plasma spray coating. Navrotsky et al.[16] showed that the C-type becomes fluorite type at 2308°C and hexagonal A-type at 2325°C. Quin et al.[17] observed for particles smaller than 10nm structural changes of yttria from C-type to B-type.

The second type of structure refers to a larger scale and is a result of fabrication method. Microstructure concerns to the quantity and distribution of the structural elements/phases in a ceramic material and determines its properties such as ionic/electric conductivity, mechanical/chemical strength, abrasion resistance, and luminescence. Microstructural formation of ceramic rods shaped by colloidal processing is shown in Fig.3. Yttrium oxide powders have as characteristic flakey particles, forming agglomerates higher than 1µm (Fig.3a$_1$). A suitable stability condition for yttrium particles and homogenization of ceramic suspensions led to formation of high compacted powders (Fig.3.a_2). The void space is eliminated by high temperature heat treatment, which results in densification of ceramic component. As shown in Fig.3a_3, as sintered yttrium oxide rods exhibited coarsened microstructure with rounded grains with size higher than 1μm.

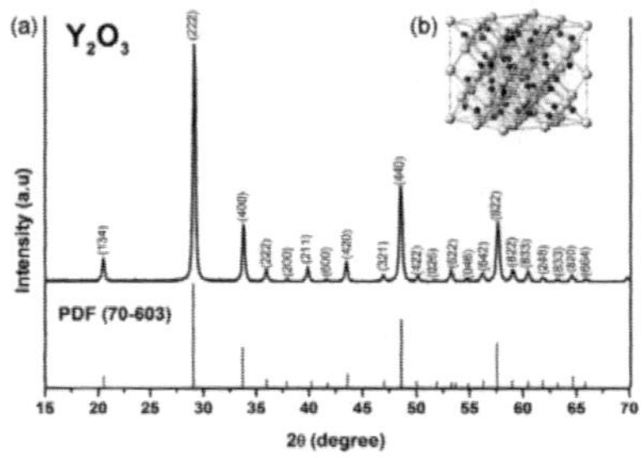

Figure 2. Crystal characterization of Y_2O_3 powders; (a) XRD curves of cubic C-type Y_2O_3; (b) model representation of crystal lattice of cubic C-type - dark spheres represent yttrium (Y) atoms and light ones oxygen (O) atoms.

Densification of a powder compact can be summarized using a model of two touching particles[22], as shown in Fig.3. Initially neck growth dominates any mass transport between particles by grain boundary diffusion (Fig.3b) i.e. particle mass diffuses along the grain boundary to the surface to form a neck and as a consequence produces shrinkage (b_1-b_2). Besides, the mass transport still continues due to the differential size of the two grains. The smaller grain donates its mass to the larger one, which reduces the surface to volume ratio. Furthermore, although neck growth ceases the two sintered grains can reduce their free energy it if the grain boundary moves through the smaller grain and disappears (b_2- b_3 and b_3-b_4). Kellet at al.[23] reported that it is achieved via coarsening where the energy barrier for grain boundary motion is zero. In the case where the smaller particle is between two larger ones (Fig. 3c) the same sequential mass transport occurs i.e sintering (c_1-c_2)> grain boundary diffusion (c_2-c_3) > sintering (c_3-c_4). Besides, it is seen that the two larger grains touch one another (c_3), initiate grain growth and reinitiate shrinkage (c_3-c_4).

Fig.4a. shows SEM image of yttrium oxide rod sintered at 1600°C for 4h at ambient atmosphere. As can be seen, as sintered ceramic samples exhibited dense microstructure (Fig.4b), grain size larger than 1μm, pycnometric density of 4.84g.cm^{-3}, which corresponds to 96,61% of theoretical density. From fracture micrograph (Fig.4c), is observed that ceramic rods exhibited dense microstructure with a closed pore situated within grain boundaries, as well as revealing transgranular fracture and cleavage planes, as indicated by white arrows.

Irradiation of materials with particles that are substantially energetic can produce atomic displacements, as well as can induce significant microstructural alteration. In contrast, gamma irradiation with doses up to 100kGy did not create microstructural changes for yttrium oxide rods. It is supposed to be ascribed to high energy bonding between Y-O atoms and the intrinsic properties of this rare earth sesquioxide. Light dots observed on grains surface (Fig.4d) are likely to be an approach of two sintered grains to reduce their free energy by moving from smaller grain to larger one.

EPR spectrum of Y_2O_3 rods irradiated with doses up to100kGy at room atmosphere and temperature are shown in Fig. 5. As it can be seen, all samples exhibited EPR signal with an isotropic g tensor of 2.00 (c_1) and line width around 23G. The main defect c_1 is ascribed to interstitial O_2^- ion, nominally addressed as superoxide ion, which is generated by adsorption of molecular oxygen as observed by Lunsford[24]. Osada et al[25] reported these interstitial O_2^- ions in a study on Y_2O_3-CaO system, with tensor value g of 2.003. Singh et. al.[26] observed that for Y_2O_3:Er powders, EPR spectra exhibited as characteristic an axially g tensor with principal values $g_{\parallel}$ of 2.0415 and $g_{\perp}$ of 2.0056. The low intensity centre (c_2) illustrated in Fig. 5 exhibits smaller width line around 13G and g tensor of 1.969. The present center is assigned to F^+ centre charged oxygen vacancy having one remaining electron. Besides, C-type crystal lattice of Y_2O_3 presents intrinsic oxygen vacancies which can produce F centres close to c_2. The effect of this point defect on luminescence of Y_2O_3 thins was discussed previously by Bordun[27]. According to his study, thermal treatment of Y_2O_3 samples in air atmosphere leads to the filling the oxygen vacancies by oxygen atoms from air into Y_2O_3 and as a result, additional weak bonds with oxygen ions are formed. On irradiation a radiative recombination between these new centres and those intrinsic centres occurs, resulting in an increase of the luminescence near 2.9eV. Singh et al.[26] evaluated the stability of signal c_2 as a function of temperature and observed that this centre decays at low temperature around 180°C.

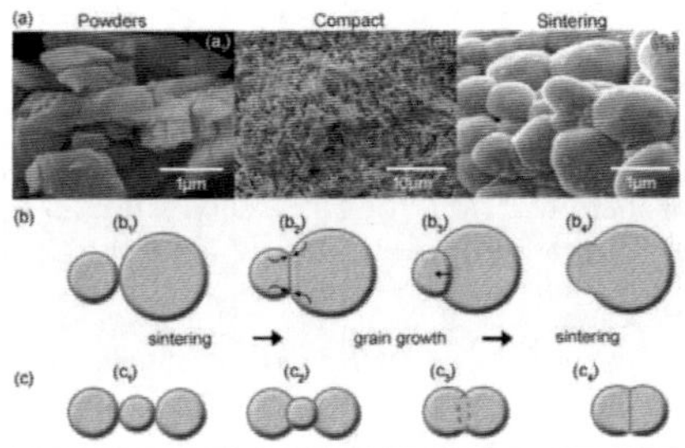

Figure 3. Sequential forming of yttrium oxide rods: (a) colloidal processing, (b) two touching particles; (c) two larger particles sandwiching a smaller particle

Figure 4. SEM images of Y_2O_3 rods sintered at 1600°C for 4h at room atmosphere, (a) ceramic rod as produced; (b) surface microstructure of unirradiated sample; (c) fracture surface, exhibiting cleavage planes in transgranular fracture (d) surface microstructure of gamma irradiated sample with 100kGy.

Fading behavior of dosimetric signal (c_1) of irradiated Y_2O_3 rods with doses up to 100kGy is illustrated in Fig. 6b. As can be seen, considering all doses the EPR relative signal of c_1 exhibited a mean decay of 23% until 96h after irradiation. Apart from this time, it became linear, revealing stability. Shivaramu et al.[31] reported that yttria samples irradiated with 6kGy gamma dose exhibited up to 22% fading over a period of sixty days.

Dose dependence of any material is the fundamental basis of its application as a dosimetric material. Within the dose range of interest the response is expected to be reproducible as a function of dose and preferably linear. Linearity describes proportionality between EPR signal and absorbed dose. Usually, the dose response behavior of most dosimetric materials exhibits linear, supralinear and saturating response.

In this work, the EPR signal as a function of gamma radiation absorbed dose, determinated as peak height of the main center (c_1), exhibited linear behavior up to 10kGy and presented a tendency of supralinearity for higher doses, as illustrated in Fig. 6a. Considering that a wide range of dose was evaluated, which was from 0,001 to 100kGy the plot scale was set to log-log in order to make visible the linearity of signal for lower doses.

Supralinearity refers to a region in which the slope of the response versus dose is greater than that for the linear region[28]. Chen et al.[29] described supralinearity as a property of the measured dose being above the continuation of the initial linear dose range. Furthermore, even if it is not related to this result, supralinearity may be ascribed to accumulation of doses, which can produce trapped electrons and holes at existing impurities along with the defects produced by the process of irradiation itself. As can be seen, Y_2O_3 rods present effective sensitivity for high dose radiation range. Salah et al.[30] as evaluating dosimetric properties of CaSO4:Dy powders observed that particles in nano scale range were more sensitive, exhibiting a linear dose response behavior for high doses. Authors suggest that this behavior and supralinearity for nanocrystalline CaSO4:Dy might be related to the probability of formation of more numbers of traps, in which a large amount of particles are exposed directly to irradiation. As a consequence, the traps will compete to fill up the electronic levels while generating or while recombining during stimulation and finally, the supralinearity will occur if there is no much retrapping.

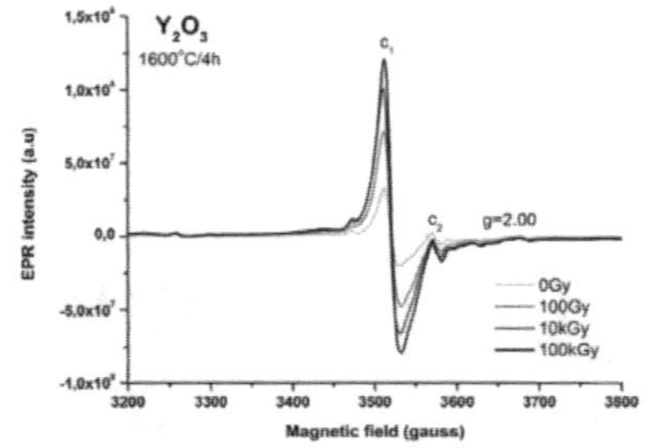

Figure 5 – EPR spectra recorded at room temperature and atmosphere for Y_2O_3 rods irradiated with doses up to 100kGy.

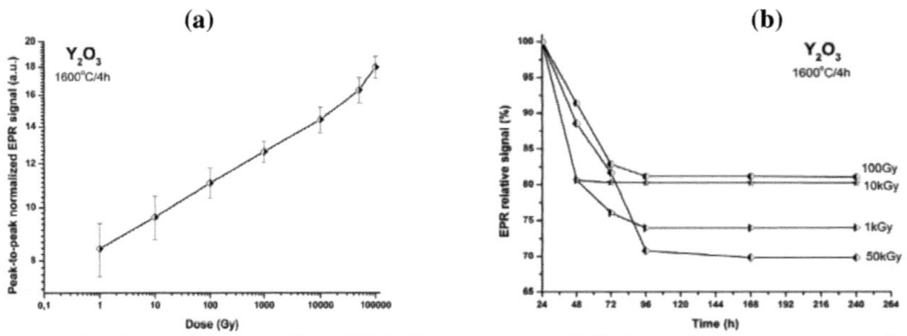

Figure 6 – (a) Peak-to-peak normalized EPR dose response of Y_2O_3 rods gamma irradiated with up to 100kGy ; (b) fading of Y_2O_3 rods at room temperature evaluated over 11 days

4. Conclusion

The effects of ionizing radiation on microstructure and electronic defects of yttrium oxide rods produced by colloidal processing followed by sintering at 1600°C for 4h in room atmosphere were evaluated. Ceramic samples exhibited dense microstructure and pycnometric density of 96.6%. None radiation induced defects were observed on microstructure of samples. However, for all samples EPR characterization revealed spectra with principal g tensor of 2.00 and maximum line width around 23G, which is ascribed to superoxide O_2^- ion generated by adsorption of molecular oxygen from room atmosphere. Dose response behaviour was linear from 0,1 to 10kGy range and supralinearity was observed for higher doses. Fading stability was achieved from 96h. The present results suggest that yttrium oxide is a promising material for high dose dosimetry applications.

5. Acknowlodgments

We authors are deeply grateful to MSc. Douglas Will Leite, MSc. William Naville, and Dr. Masao Matsuoka. In addition, grant #2014/23621-3, São Paulo Research Foundation (FAPESP); National Council for Scientific and Technological Development (CNPq); and Coordination for Improvement of High Degree People (CAPES).

6. References

[1] F.F. Li, L.Q. Xiao, L.J. Liu, Polymeric palladium-mediated carbene polymerization, Polym Chem-Uk, 6 (2015) 6163-6170.
[2] R.R. Cheng, Z.L. Wu, Y.L. Hou, J. Dong, J.Z. Cui, B. Zhao, Three Cu(II) coordination polymers with novel bi-triazole ligand: Synthesis, structure and EPR properties, Inorg Chem Commun, 51 (2015) 95-98.
[3] N.E. Domracheva, V.E. Vorobeva, M.S. Gruzdev, A.V. Pyataev, Blue shift in optical absorption, magnetism and light-induced superparamagnetism in gamma-Fe2O3 nanoparticles formed in dendrimer, J Nanopart Res, 17 (2015).

[4] M. Hashem, M. Weiler-Sagie, P. Kuppusamy, G. Neufeld, M. Neeman, A. Blank, Electron spin resonance microscopic imaging of oxygen concentration in cancer spheroids, J Magn Reson, 256 (2015) 77-85.
[5] M.A. Morsy, A.N.M. Kawde, Electron paramagnetic resonance monitoring for on-demand electrochemically-generated radicals, Electrochim Acta, 160 (2015) 22-27.
[6] M. Marina, M.Z.M. Zamzuri, M.N. Derman, M.A. Selamat, Z. Nooraizedfiza, Comparison Study in Consolidation of Yttria Reinforced Iron-Chromium Composites using Conventional and Microwave Sintering Technique, Advanced Materials Engineering and Technology Ii, 594-595 (2014) 832-836.
[7] F. Hayashi, M. Tanaka, D.M. Lin, M. Iwamoto, Surface structure of yttrium-modified ceria catalysts and reaction pathways from ethanol to propene, J Catal, 316 (2014) 112-120.
[8] X.M. Han, X. Feng, X.W. Qi, X.Q. Wang, M.Y. Li, The Photoluminescent Properties of Y2O3:Bi3+, Eu3+, Dy3+ Phosphors for White-Light-Emitting Diodes, J Nanosci Nanotechno, 14 (2014) 3387-3390.
[9] A. Jyotsana, G.S. Maurya, A.K. Srivastava, A.K. Rai, B.K. Ghosh, Synthesis and electrical properties of Y2O3: Dy3+ & Eu3+ nanoparticles, Appl Phys a-Mater, 117 (2014) 1269-1274.
[10] T. Mongstad, A. Thogersen, A. Subrahmanyam, S. Karazhanov, The electronic state of thin films of yttrium, yttrium hydrides and yttrium oxide, Sol Energ Mat Sol C, 128 (2014) 270-274.
[11] M.A. Auger, V. de Castro, T. Leguey, J. Tarcisio-Costa, M.A. Monge, A. Munoz, R. Pareja, Effect of yttrium addition on the microstructure and mechanical properties of ODS RAF steels, J Nucl Mater, 455 (2014) 600-604.
[12] J. Hostasa, J. Matejicek, B. Nait-Ali, D.S. Smith, W. Pabst, L. Esposito, Thermal Properties of Transparent Yb-Doped YAG Ceramics at Elevated Temperatures, J Am Ceram Soc, 97 (2014) 2602-2606.
[13] S.C. Santos, W. Acchar, C. Yamagata, S. Mello-Castanho, Yttria nettings by colloidal processing, J Eur Ceram Soc, 34 (2014) 2509-2517.
[14] H.R. Hoekstra, K.A. Gingerich, High-Pressure B-Type Polymorphs of Some Rare-Earth Sesquioxides, Science, 146 (1964) 1163-&.
[15] V. Gourlaouen, G. Schnedecker, A.M. Lejus, M. Boncoeur, R. Collongues, Metastable Phases in Yttrium-Oxide Plasma Spray Deposits and Their Effect on Coating Properties, Mater Res Bull, 28 (1993) 415-425.
[16] A. Navrotsky, L. Benoist, H. Lefebvre, Direct calorimetric measurement of enthalpies of phase transitions at 2000 degrees-2400 degrees C in yttria and zirconia, J Am Ceram Soc, 88 (2005) 2942-2944.
[17] X. Qin, Y.G. Ju, S. Bernhard, N. Yao, Flame synthesis of Y2O3 : Eu nanophosphors using ethanol as precursor solvents, J Mater Res, 20 (2005) 2960-2968.
[18] M.F. Yan, Effects of physical, chemical, and kinetic factors on ceramic sintering, in: K.S.M. G.L. Messing, J.W. McCauley, R. A. Haber (Ed.) Advanced Ceramics, American Ceramic Society, Westerville, OH, 1987.
[19] M.A. Yar, S. Wahlberg, M.O. Abuelnaga, M. Johnsson, M. Muhammed, Processing and sintering of yttrium-doped tungsten oxide nanopowders to tungsten-based composites, J Mater Sci, 49 (2014) 5703-5713.
[20] X.B. Li, Q.Q. Zhang, S.Y. Ma, G.X. Wan, F.M. Li, X.L. Xu, Microstructure optimization and gas sensing improvement of ZnO spherical structure through yttrium doping, Sensor Actuat B-Chem, 195 (2014) 526-533.
[21] V. Kumar, O.P. Pandey, K. Singh, K. Lu, Interaction Study of Yttria-Based Glasses with High-Temperature Electrolyte for SOFC, Fuel Cells, 14 (2014) 635-644.
[22] F.F. Lange, B.J. Kellett*, Thermodynamics of Densification: II, Grain Growth in Porous Compacts and Relation to Densification, J Am Ceram Soc, 72 (1989) 735-741.
[23] B.J. Kellett, F.F. Lange, Thermodynamics of Densification: I, Sintering of Simple Particle Arrays, Equilibrium Configurations, Pore Stability, and Shrinkage, J Am Ceram Soc, 72 (1989) 725-734.
[24] J.H. Lunsford, Esr of Adsorbed Oxygen Species, Catal Rev, 8 (1973) 135-157.
[25] Y. Osada, S. Koike, T. Fukushima, S. Ogasawara, T. Shikada, T. Ikariya, Oxidative Coupling of Methane over Y2o3-Cao Catalysts, Appl Catal, 59 (1990) 59-74.
[26] V. Singh, V.K. Rai, I. Ledoux-Rak, S. Watanabe, T.K.G. Rao, J.F.D. Chubaci, L. Badie, F. Pelle, S. Ivanova, NIR to visible up-conversion, infrared luminescence, thermoluminescence and defect centres in Y(2)O(3) : Er phosphor, J Phys D Appl Phys, 42 (2009).
[27] O.M. Bordun, Influence of Oxygen Vacancies on the Luminescence Spectra of Y2O3 Thin Films, J Appl Spectrosc+, 69 (2002) 430-433.
[28] A. Standard, Standard practice for use of thermoluminescence dosimetry (TLD) systems for radiation processing, 2005.
[29] R. Chen, S.W.S. Mckeever, Characterization of Nonlinearities in the Dose Dependence of Thermoluminescence, Radiat Meas, 23 (1994) 667-673.
[30] N. Salah, P.D. Sahare, S.P. Lochab, P. Kumar, TL and PL studies on CaSO4: Dy nanoparticles, Radiat Meas, 41 (2006) 40-47.

[31] N.J. Shivaramu, B.N. Lakshminarasappa, K.R. Nagabhushana, F. Singh, Synthesis characterization and luminescence studies of gamma irradiated nanocrystalline yttrium oxide, Spectrochim Acta A, 154 (2016) 220-231.

A deeper insight into (Lu,Y)AG:Pr scintillator crystals

W. Drozdowski[1], A.J. Wojtowicz[1], K. Brylew[1], W. Łachmański[1], E. Talik[2], M. Szubka[2], J. Kusz[2], A. Guzik[2], K. Balin[3], J. Kisielewski[4], M. Świrkowicz[4] and A. Pajączkowska[4]

[1]Institute of Physics, Faculty of Physics, Astronomy and Informatics, Nicolaus Copernicus University, Grudziadzka 5, 87-100 Torun, Poland
[2]Institute of Physics, University of Silesia, Uniwersytecka 4, 40-007 Katowice, Poland
[3]Silesian Center for Education and Interdisciplinary Research, University of Silesia, 75 Pulku Piechoty 1A, 41-500 Chorzow, Poland
[4]Institute of Electronic Materials Technology, Wolczynska 133, 01-919 Warsaw, Poland

E-mail: wind@fizyka.umk.pl

Abstract. Interior of Czochralski-grown (Lu,Y)AG:Pr crystals has been examined by means of several techniques, such as X-Ray Photoelectron Spectroscopy, X-Ray Diffraction, Time-of-Flight Secondary Ion Mass Spectrometry, and magnetic susceptibility measurements. Additionally, their luminescence has been monitored at various combinations of a double-beam (X-ray/IR) excitation.

1. Introduction

Mixed (Lu,Y)AG:Pr crystals have already been recognized as promising fast and efficient scintillator materials. Compared to their prototype LuAG:Pr they offer higher light outputs and increased contributions of the prompt component to their scintillation time profiles at an expense of a somewhat lower density [1,2]. In this Communication we provide additional data on the internal structure of Czochralski-grown (Lu,Y)AG:Pr crystals, as well as on some trap-related aspects of their scintillation mechanism. Our research comprises X-Ray Photoelectron Spectroscopy (XPS), X-Ray Diffraction (XRD), Time-of-Flight Secondary Ion Mass Spectrometry (ToF-SIMS), measurements of magnetic susceptibility as a function of temperature, and advanced experiments with a coupled X-ray and infrared (IR) excitation.

2. Materials and experiment

The crystals of pure and Pr-activated $(Lu_xY_{1-x})_3Al_5O_{12}$ (x = 0.00, 0.25, 0.50, 0.75, 1.00) have been grown at ITME, Warsaw, by the Czochralski method as characterized in [1,2].

XPS spectra have been taken with a PHI 5700/660 Physical Electronics Photoelectron Spectrometer under monochromated Al K_α X-rays (14.86 keV). XRD has been investigated with a four-circle Kuma Diffraction KM-4 diffractometer under graphite-monochromated Mo K_α X-rays (17.48 keV). ToF-SIMS studies have been carried out with a IONTOF TOF.SIMS 5 reflection-type spectrometer equipped with a Bi liquid metal ion gun. The data analysis has been performed with a

SurfaceLab6 software package. Magnetic susceptibility has been measured as a function of temperature between 2 and 400 K using a Quantum Design MPMS-XL-7AC SQUID magnetometer.

Double-beam experiments have been conducted with a setup consisting of an Inel XRG3500 X-ray generator with a Cu-anode tube (45 kV, 10 mA), a Thorlabs IR laser diode (830 nm, 1000 mW, 750 mA), an ARC SpectraPro 500i monochromator, a Hamamatsu R928 photomultiplier, and an APD Cryogenics closed-cycle He cooler with a Lake Shore 330 temperature controller.

3. Results and discussion

3.1. Internal structure studies

3.1.1. XPS. Two representative XPS spectra of $(Lu_xY_{1-x})_3Al_5O_{12}$:Pr ($x = 0.75$ and 0.25) are shown in Fig. 1. They are clearly dominated by well separated lines related to the crystal constituents, i.e. Lu, Y, Al, and O [3,4]. Besides, there is only a single weak line attributed to none of the above-mentioned elements. It is the C 1*s* line at 285 eV, arising in XPS of most samples due to their surface contamination with trace amounts of C, which is in fact useful for calibration purposes [5]. Based on their XPS spectra, the quality of the (Lu,Y)AG:Pr crystals may be rated very high.

3.1.2. XRD. The lattice parameters (*LP*) of $(Lu_xY_{1-x})_3Al_5O_{12}$:Pr read out from XRD measurements are specified in Table 1 and depicted in Fig. 2 as a function of *x*. It can easily be noticed that *LP* decreases linearly with *x*, which correlates with the smaller ionic radius of Lu^{3+} compared to Y^{3+} [6] and satisfies the Vegard's law [7]. We note that the values of both the slope ($m = 0.091$ Å) and the intercept ($LP_0 = 12.01$ Å) derived from a straight line fit:

$$LP = -mx + LP_0 \tag{1}$$

are in a full accordance with the data of Kuwano *et al.* [8]. Since the particular *LP* points do not lie perfectly on this line, there must be some discrepancies between the initial (in the melt) and the real Lu-to-Y ratios in the mixed crystals, nevertheless their magnitude is very small.

3.1.3. ToF-SIMS. The performed surface imaging and depth profiling, providing information on the chemical composition of the (Lu,Y)AG:Pr crystals, confirm their high purity. A representative case is illustrated by Fig. 3, containing a depth profile of LuAG:Pr. The spatial distribution of the crystal constituents (Lu, Al, O, Pr) can be regarded as uniform. Although some other elementary (H, Li, C, Na, Mg, Si, K, Ca, Ti, Fe, Cu, Zn) and molecular ions appear in the recorded mass spectra, the amount of such contaminations remains at a standard low level, strongly decreasing with depth.

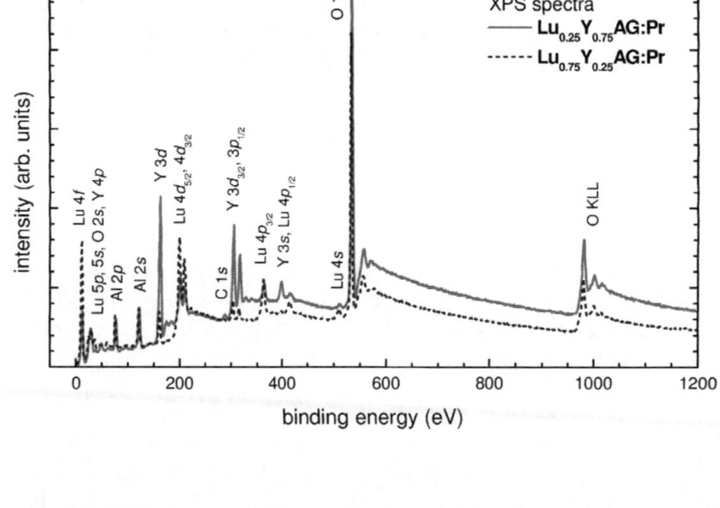

Figure 1. XPS spectra of $Lu_{0.75}Y_{0.25}$AG:Pr and $Lu_{0.25}Y_{0.75}$AG:Pr.

Table 1. Parameters of the studied crystals derived from the XRD and magnetic susceptibility measurements. For a comparison, Pr concentrations determined by the ICP-OES [1] are also listed.

Crystal	LP (Å)	χ_0 (10^{-3} emu/mol)	θ(K)	C (10^{-3} emu·K/mol)	Pr concentration (at%) magn. susc.	Pr concentration (at%) ICP-OES
LuAG	11.917	-0.244	-	-	-	-
LuAG:Pr	11.914	-0.231	-8.3	5.19	0.33	0.12
$Lu_{0.75}Y_{0.25}AG$:Pr	11.940	-0.211	-7.9	5.20	0.33	0.16
$Lu_{0.5}Y_{0.5}AG$:Pr	11.959	-0.230	-11.0	7.71	0.48	0.17
$Lu_{0.25}Y_{0.75}AG$:Pr	11.986	-0.150	-8.0	5.71	0.36	0.23
YAG	12.005	-0.208	-	-	-	-

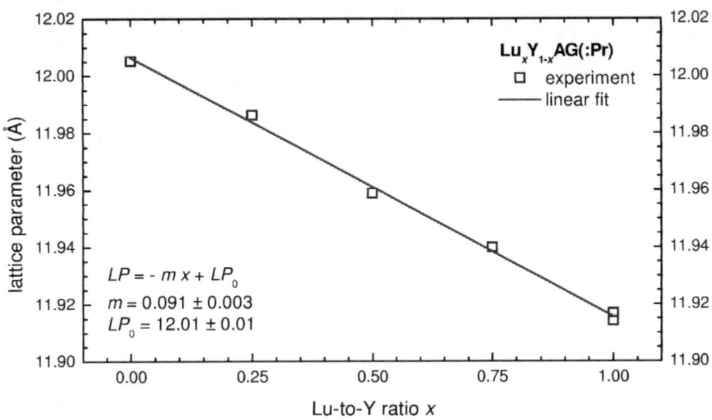

Figure 2. Lattice parameter of $(Lu_xY_{1-x})_3Al_5O_{12}$ as a function of x.

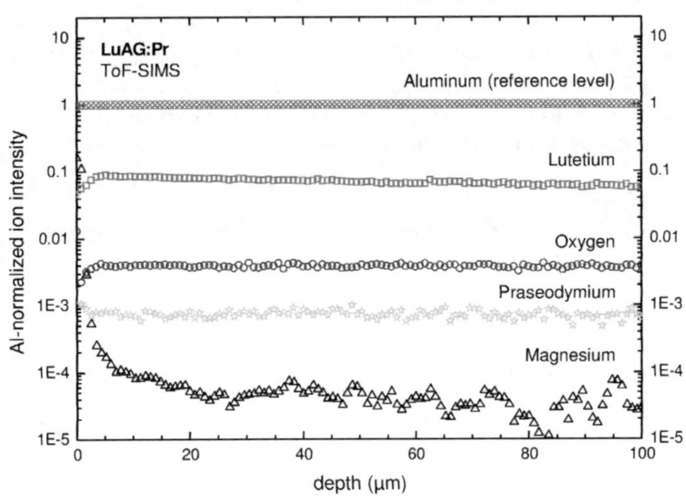

Figure 3. A ToF-SIMS depth profile of LuAG:Pr. Intensities corresponding to the crystal constituents (Lu, O, Pr) and an exemplary contamination (Mg) are normalized to the level of Al.

3.1.4. Magnetic susceptibility. To verify the Pr content determined before by the ICP-OES technique [1], the magnetic susceptibilities χ of the (Lu,Y)AG:Pr crystals have been measured as a function of temperature T (Fig. 4) and thereafter fitted to the known Curie–Weiss law:

$$\chi = \frac{C}{T-\theta} + \chi_0 \quad (2)$$

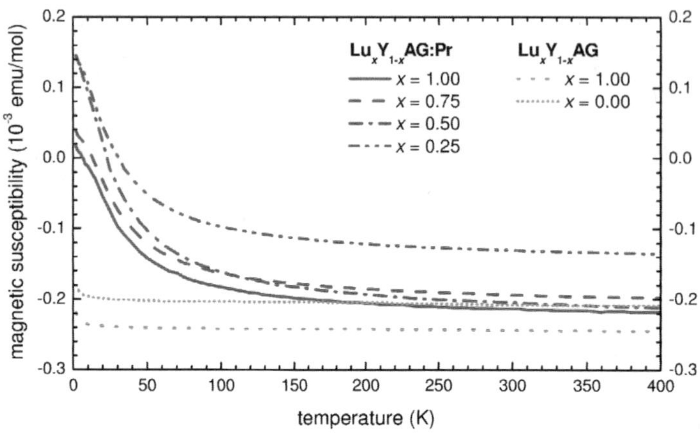

Figure 4. Magnetic susceptibility of (Lu,Y)AG:Pr and (Lu,Y)AG as a function of temperature.

between 50 and 400 K, where *C* is the Curie constant, θ - the Curie–Weiss temperature, and χ_0 - the temperature-independent diamagnetic susceptibility. The fitting parameters, together with the estimated Pr concentrations using the approach of Talik *et al.* [5], are listed in Table 1. Surprisingly, the obtained concentration values are 2-3 times higher and there is hardly any correlation with the ICP-OES results, which clearly needs a further investigation.

3.2. Double-beam experiments

We have used a novel method proposed by Poolton *et al.* [9] to study the role of traps in the scintillation of (Lu,Y)AG:Pr. During the experiment the luminescence intensity is recorded while the sample is excited by X-rays, X-rays and IR radiation simultaneously, and by IR alone. Both sources are operated independently, but in a sequence to monitor X-ray excited radioluminescence (RL), IR-stimulated luminescence (IR-SL), and RL + IR-SL altogether (Fig. 5). We observe that a longer X-ray irradiation with no IR stimulation leads to a luminescence decay (after switching X-rays off) showing a fast (direct) component and some larger contribution from slow components (afterglow). The time scale of the experiment does not allow any detection of the fast component (radiative lifetime) that is most likely hidden in the direct component. No dominant single exponential decay is present in the IR-SL decays (with or without X-ray excitation). A shorter or no X-ray irradiation is

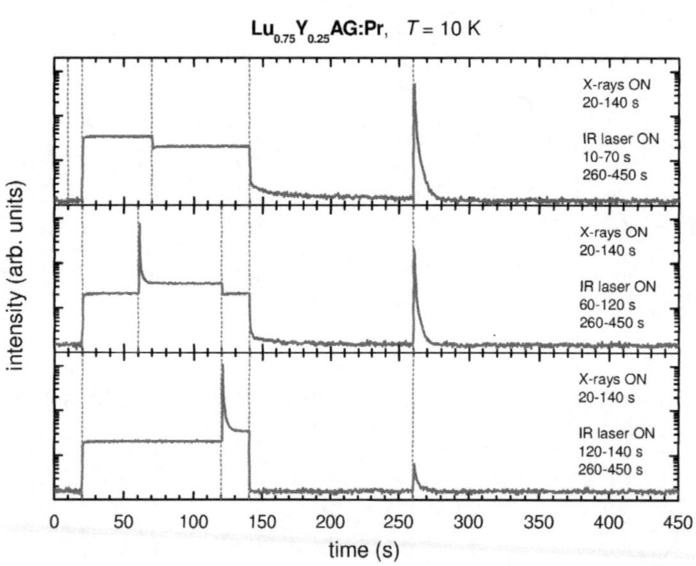

Figure 5. Intensity of the X-ray, IR laser, and X + IR excited luminescence of $Lu_{0.25}Y_{0.75}AG$:Pr at 10 K vs. time. The emission wavelength is set at 365 nm. The timing and sequence of excitations is indicated in each panel. Note the log scale.

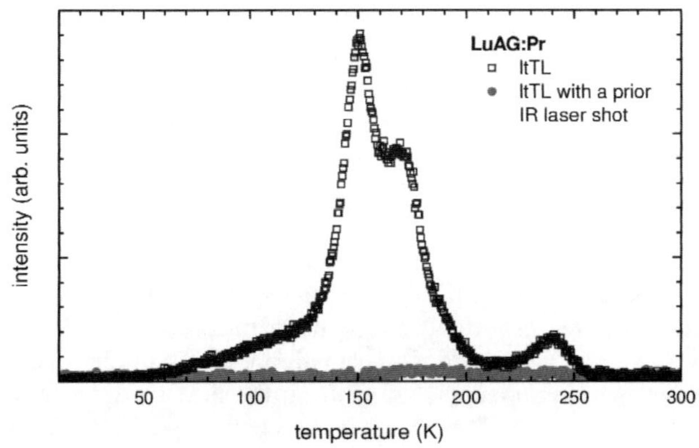

Figure 6. Glow curves of LuAG:Pr, recorded at a heating rate of 0.14 K/s after a standard X-ray irradiation (10 min), with and without a preceding IR laser shot (1 s).

also responsible for a less efficient stimulation by IR, which is consistent with low temperature thermoluminescence (ltTL) measurements: the glow curves in Fig. 6 reveal that all ltTL glow peaks below 250 K are strongly reduced, pointing to traps that contribute to the IR-SL and are emptied by the 830 nm laser irradiation.

It is well known that recombination kinetics can usually be described by a bimolecular law [10]:

$$N = An^2 = \frac{An_0^2}{(n_0At+1)^2} \tag{3}$$

where N stands for the number of photons emitted from unit volume per unit time, n – the number of free electrons per unit volume available for recombination, here equal to the number of Pr^{4+} ions [11], whilst the probability that within the time t a given electron will recombine with one of the n available Pr^{4+} ions is equal to Ant. Plotting then:

$$\frac{1}{\sqrt{I}} = \frac{1}{\sqrt{I_0}}(1 + An_0t) \tag{4}$$

we obtain a decay represented by a straight line.

To check this idea, in Fig. 7 we replot the middle panel of Fig. 5 using the appropriate scale (1/sqrt(intensity)). It is interesting to note that the IR-SL decay, measured after the IR laser is turned on (at 260 s), can be fitted reasonably well with a straight line labeled (a), thus following closely a bimolecular decay law. This is to be expected assuming that *i)* a prior X-ray irradiation fills the traps

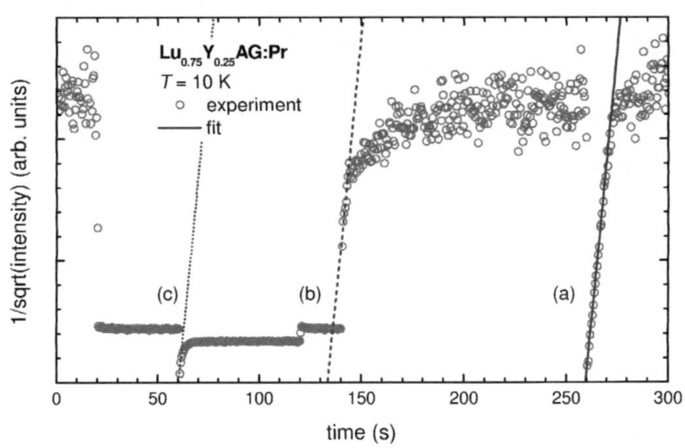

Figure 7. Intensity of the X-ray, IR laser, and X + IR excited luminescence of $Lu_{0.25}Y_{0.75}AG$:Pr at 10 K vs. time. The emission wavelength is set at 365 nm. The solid (a) straight line is a linear fit to selected experimental points of the IR-SL decay. The dashed (b) and dotted (c) straight lines have the same slope but are shifted to fit different decays. Note the inverse square root scale.

with electrons and generates some population of Pr^{4+} ions, *ii)* electrons are released from traps upon the IR irradiation, *iii)* there is no retrapping of electrons, and *iv)* there is no significant trapping by deeper traps (so that the concentration of Pr^{4+} is not much higher than the number of electrons ready to recombine). The two remaining straight lines in Fig. 7, labeled (b) and (c), have the same slope, but have been shifted to fit the experimental points corresponding to the decays observed when the X-ray and IR excitations are simultaneously turned off (b) or when the IR excitation is turned off but the X-ray excitation remains at a fixed value (c).

It can be noticed in Fig. 7 that unlike for the IR laser stimulation, the decay (b) after the X-ray and IR excitations are turned off follows neither a simple bimolecular law nor a single exponential decay. The experiment, limited in the time resolution, proves nevertheless that there must be an additional component that dominates the shorter times. A simple and obvious explanation is by assuming that in the decay following the X-ray excitation there is a component coming from the direct excitation of Pr^{3+} ions. No such effect is observed in the third decay (c). This decay is dominated by a bimolecular component imposed on the steady-state signal coming from the X-ray excitation.

4. Conclusions

The internal structure research generally confirms a high quality of the grown (Lu,Y)AG:Pr crystals. The only source of concern is the discrepancy between the evaluated Pr concentrations based on two different methods. This issue deserves a further interest.

A bimolecular decay of the X-ray excited sample stimulated by the IR radiation, perceived in the double-beam studies, directly proves that electron transport over the (Lu,Y)AG host is possible. Although the contribution of various processes to scintillation decays may be different than observed in such a kind of a steady-state experiment, the possibility of a bimolecular recombination at activator sites in scintillators should not be overlooked. Finally, since there is no indication of any excitonic transport, the main scintillation mechanism in (Lu,Y)AG:Pr must be a consecutive recombination of first holes and then electrons at Pr sites, which agrees with the previous assumptions [11].

5. References

[1] Drozdowski W, Brylew K, Wojtowicz A J, Kisielewski J, Świrkowicz M, Łukasiewicz T, de Haas J T M and Dorenbos P 2014 *Opt. Mater. Express* **4** 1207-12

[2] Drozdowski W, Brylew K, Witkowski M E, Drewniak A, Masewicz Z, Wojtowicz A J, Kisielewski J and Świrkowicz M 2016 *Opt. Mater.* **59** 107-14

[3] Pawlak D A, Woźniak K, Frukacz Z, Barr T L, Fiorentino D and Seal S 1999 *J. Phys. Chem.* B **103** 1454-61

[4] Kruczek M, Talik E, Sakowska H, Gała M and Świrkowicz M 2005 *Cryst. Res. Technol.* **40** 439-43

[5] Talik E, Zajdel P, Guzik A, Skrzypek D, Lipińska L and Michalska M 2014 *J. Alloys Compd.* **616** 556-68

[6] Shannon R D 1976 *Acta Cryst.* A **32** 751-67

[7] Vegard L 1921 *Z. Phys.* **5** 17-26

[8] Kuwano Y, Suda K, Ishizawa N and Yamada T 2004 *J. Cryst. Growth* **260** 159-65

[9] Poolton N R J, Bos A J J, Wallinga J, de Haas J T M, Dorenbos P, de Vries L, Kars R H, Jones G O and Drozdowski W 2010 *J. Lumin.* **130** 1404-14

[10] Mott N F and Gurney R W 1948 *Electronic processes in ionic crystals* (Oxford: Clarendon Press) p 209

[11] Drozdowski W, Dorenbos P, de Haas J T M, Drozdowska R, Owens A, Kamada K, Tsutsumi K, Usuki Y, Yanagida T and Yoshikawa A 2008 *IEEE Trans. Nucl. Sci.* **55** 2420-4

Acknowledgements

This work has been financed from the funds of the Polish National Science Centre granted on the basis of Decisions no. DEC 2012/05/B/ST5/00324 and DEC 2013/09/N/ST3/01935.

Quenching factor for alpha particles in ZnSe scintillating bolometers

S Nagorny[1,2], L Cardani[3,4], N Casali[3,5], I Dafinei[3], L Pagnanini[1], L Pattavina[2], S Pirro[2] and K Schaeffner[1,2]

[1] Gran Sasso Science Institute, I-67100 L'Aquila – Italy
[2]INFN - Laboratori Nazionali del Gran Sasso, I-67100 Assergi - Italy
[3] INFN - Sezione di Roma I, I-00185 Roma - Italy
[4]Physics Department - Princeton University, 08544 Princeton - USA
[5]Dipartimento di Fisica - Università di Roma La Sapienza, I-00185 Roma - Italy

E-mail: sergey.nagorny@lngs.infn.it

Abstract. In the framework of the CUPID-0 experiment, a numbers of ZnSe single crystals were produced and subjected to different thermal treatments, and later tested as cryogenic scintillating bolometers. We have found that a specific thermal treatment (24 hours under argon atmosphere at 900 °C) has a strong impact on some properties of ZnSe crystals (amplitude of signal, light yield, specific resistivity) and most interestingly, changes the quenching factor for alpha particles from values > 1 to values < 1. Thus such thermal treatment opens the possibility to modify this experimental parameter for a various applications.

1. Introduction

Cryogenic scintillating bolometers are a promising technology to explore diverse rare nuclear processes like rare alpha decay, rare or forbidden beta decays, neutrinoless double beta decay, which have characteristic decay times larger than 10^{14} y. The main feature of scintillating bolometers is the simultaneous heat release and light emission caused an interacting particle in the crystal. Read-out of both these signals allows for an active discrimination of the type of interacting particle since the amount of emitted light depends on the nature of the particle. Besides excellent particle discrimination, the bolometers also possess a very good energy resolution and nearly 100% detection efficiency.

Observations including ^{209}Bi alpha decay ($T_{1/2} = 1.9\times10^{19}$ y) [1], ^{180}W alpha decay (1.8×10^{18} y) [2], ^{151}Eu alpha decay (4.6×10^{18} y) [3] and ^{148}Sm alpha decay (4.6×10^{15} y) [4] are good examples to power for this technique. ZnSe scintillating bolometers are the fundamental technology of the CUPID-0 experiment which is searching for neutrinoless double beta decay of ^{82}Se.

Zinc selenide, especially when doped with Te or Al, has a long history of use [5] particularly as X-ray and gamma-ray detectors [6]. However, ZnSe crystals have an interesting feature: alpha particles produce more light than electrons at the same deposited energy in the detector, i.e. Quenching Factor (QF) is greater than unity. To our knowledge no other scintillator with a QF for alphas greater than 1 has ever been reported in literature. Such anomalous behavior of scintillation cannot be easily accommodated within the theoretical framework used to describe the scintillation properties of materials. Indeed, heavy particles usually have a QF smaller than one and it is explained on the basis of a saturation effect due to the high ionization density that characterizes the interaction of heavy particle in matter [7]. Although this feature is still not understood it is very important for particle discrimination. When the light yield is a strong function of the particle type, the effective particle discrimination can be achieved.

The results reported here reflect ongoing R&D activities within the CUPID-0 experiment to improve the ZnSe scintillating bolometers performance. We have investigated methods to reduce after-growth thermal stress which can trigger cracking of the crystals. We have found that the anomalous QF for alpha particles in ZnSe crystals depends on certain type of defects and it can be adjusted by thermal treatment.

2. Scintillating bolometers and experimental details

The working principle of bolometers is based on measuring the temperature rise induced by particle interactions in a crystal absorber that is operated at mK-temperatures. The interaction in the scintillating crystal produces ionization and excitation, triggering a series of processes that lead to the conversion of largest part of the deposited energy into phonons (i.e. lattice elementary excitations). Some fraction of the energy may be avoiding phonon conversion, for example populating long-lived excited states of defects or other *hidden* energy channels. In scintillating bolometers, some energy is transferred to luminescence centers and the emitted photons escape the crystal. In this case, for each particle interaction two signals are recorded: a phonon signal produced in the scintillating crystal and a light signal read-out by an adjacent Light Detector (LD). The ratio between the two (light-to-heat) depends on the particle Light Yield (LY) that is defined here as the fraction of the total energy deposited in the crystal, which is converted into the light. Betas and gammas show the same LY, which is typically different from the LY of alpha particles. Consequently, the simultaneous read-out of heat and light signal allows for particle discrimination.

The temperature rise induced by the particle interaction in the scintillating crystals and by the scintillation photons in the LD are read out by means of Germanium Neutron Transmutation Doped thermistors (Ge-NTD) [8]. The Ge-NTDs are coupled to their respective absorbers using a small amount of epoxy resin. In order to monitor the drift of thermal gain of the ZnSe bolometer and LD, a heavily doped silicon resistor was glued on the respective detectors to inject calibrated heat pulses. Temperature drifts can thus be corrected off-line on the basis of the measured thermal gain variation [9,10]. The crystal and the LD are housed in a copper structure similar to the one described in [11] and kept in position by PTFE tips. The ZnSe crystals are surrounded by a reflecting foil (3M VM2002) to increase the overall light collection efficiency; however they are not in thermal contact with the foil. A high purity Ge wafer (∅44.5×0.2 mm) is used as a LD, more details can be found in [12].

Ge-NTD thermistors transduce the energy deposits in the crystal and the light detector into voltage signals which are amplified and fed into an 18-bit ADC. The entire waveform of each voltage signal is amplified and continuously sampled and digitized by an ADC. At the ADC input, a low pass Bessel filter is used [13]. The trigger of the ZnSe crystal was software generated, while the LD is independently acquired in coincidence with the former. . The waveforms are filtered with the optimal filter technique [13,14] to improve the signal-to-noise performance and the amplitude (V_{heat}), taken as the maximum of the filtered pulse, is determined.

The ZnSe bolometers were energy calibrated with external gamma sources (^{40}K and ^{232}Th) placed outside the cryostat. Gamma lines between 911.2 keV and 2614.5 keV were used to determine the amplitude (V_{heat}) to energy (E) conversion function of the heat signal, attributing to each identified peak the nominal energy of the gamma line as if all the energy was converted into heat. Since it is performed using the electrons produced by gamma rays, it is commonly indicated as "electron equivalent" (keVee). The energy dependence of amplitude was parameterized with a second order polynomial with zero intercept in log (V_{heat}).

In order to compare the detector response to alpha and beta/gamma particles the ZnSe crystals were exposed in all runs to an external low-intensity smeared alpha ^{238}U source, which provides alpha particles in range from 0 to 4.5 MeV.

A weak ^{55}Fe source was permanently facing the LD, on the side opposite to the ZnSe crystal scintillator, to allow for a direct energy calibration of the LD. The signal amplitude from the light energy was assumed to have a linear dependence in the range of 0 – 35 keV. An energy resolution (FWHM) of 590 eV was evaluated for the X-ray doublet of 5.9 and 6.5 keV. The ^{55}Fe calibration provides a method to measure and compare the LY of different type of particles, but it cannot be used for an absolute evaluation of the amount of scintillation light emitted since the light amplitude is not corrected for the photon collection efficiency.

3. Crystals and thermal treatment

For our studies we have used three different ZnSe crystals whose characteristics are reported in table 1. The crystals have been produced by the high pressure Bridgman technique in a graphite crucible under argon atmosphere in the Institute for Scintillation Materials of NASU (Kharkov, Ukraine). Two of the crystals are nominally undoped, while the third one is doped by Al at the level of 10 ppm. Their initial color variation is caused by slightly different initial stoichiometry – crystals #1 and #2 were produced from ZnSe powder, while crystal #3 was produced from Chemical Vapor Deposition (CVD) smashed pieces - and the thermal conditions during crystal growth, that have also a final impact on stoichiometry of the melt [15], as well as on defect types and amount.

Table 1. Characteristics of the three different used ZnSe crystals.

	#1	#2	#3
Dimensions, mm	∅43×27	∅43×44	∅45×50
Mass, g	250	329	430
Initial charge	ZnSe powder	ZnSe powder	CVD pieces
Doping	undoped	undoped	Al (10 ppm)

For the thermal treatment the crystals were placed in a graphite crucible, which then was placed in a quartz ampule. The quartz container was put under a low pressure of argon (99.93% purity) and placed in a vertical oven, which was heated up to 900°C (100°C/h). The annealing was done at 900°C for 24 hours followed by cooling down at a rate of 100°C/h. After this annealing, for all three crystals were undertaken a cryogenic test as scintillating bolometers.

4. Effect of thermal treatment on detector performance

The cryogenic tests have been performed operating the ZnSe crystals as scintillating bolometers at about 15 mK, in a low temperature refrigerator installed underground in Hall C of Laboratori Nazionali del Gran Sasso (LNGS), Italy. In all runs the ZnSe bolometers worked properly but they cooled more slowly than all other crystals ever tested.

The thermal treatment (TT) significantly reduced the signal amplitude, computed as the maximum of the filtered pulse, both in the heat and the light channel for all crystals under investigation. Table 2 gives a summary on the recorded signal amplitude before and after TT.

Table 2. Impact of thermal treatment on the signal's amplitude in the heat and the light channel of ZnSe scintillating bolometers. Values of the signal amplitudes are expressed in μV/MeV.

	#1		#2		#3	
	heat	light	heat	light	heat	light
Before TT	6.5	1700	6.5	400	50	980
After TT	1.5	250	5.0	210	13	310

Such a significant decrease in signal amplitude is about five times in case of the ZnSe #1 and is about three times in case of the ZnSe #3 crystal cannot be explained by the different mounting of the thermistors before and after TT. The observation suggests changes of the crystal defect structure and the appearance of a large amount of high mobility charge carriers. Last conclusion is supported by

IOP Conf. Series: Materials Science and Engineering **169** (2017) 012011 doi:10.1088/1757-899X/169/1/012011

measurements of the specific resistivity of the ZnSe material. In case of the ZnSe #3, the crystal's specific resistivity decreased from 10^{10} Ohm·cm to 10^{7} Ohm·cm after TT. Since the material resistivity is proportional to the inverse product of the number of defects and the mobility of charge carriers, the TT drastically affects on both parameters. At the same time, the TT decreases also the number of luminescence centers or changes their nature, which results in a reduction of the light signals by a factor between 2 and 5 (see table 2).

After TT the heat capacity of the crystals increased, indicating an increase in the crystal lattice distortion due to introduction of new defects and/or new types of defects. Evidence for changes of the crystal structure is also derived from the light and thermal signal pulse-shape.

The time response of the LD is extremely slow – the light pulse has a rise time of the order of a few ms and decay time of the order of hundreds ms. However, at low temperature, also the light emission from scintillators is characterized by a long time constants, with different values for alphas or beta/gammas [[16]]. This is true for ZnSe crystals and thus a discrimination based on a pulse shape analysis is achievable. The shape of heat pulses also appears to be different for alpha and beta/gamma events [[17]]. As can be seen from figure 1, the TT changed the kinetics of luminescence; the radiative recombination of luminescent centers is demonstrating redistribution among different components of scintillation signal.

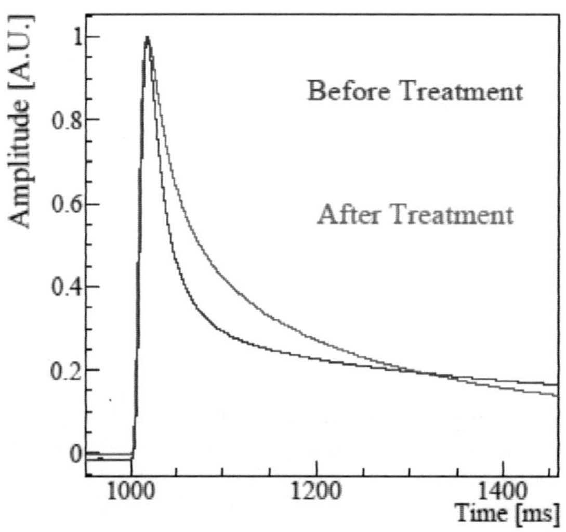

Figure 1. Impact of thermal treatment on the pulse-shape of the light signal for the ZnSe #2 (Color online).

Alpha vs. beta/gamma discrimination in scintillating bolometers may be visualized using light vs. heat scatter plot. In such a plot, alphas and betas/gammas populate separate bands by virtue of their differing light to heat ratio.

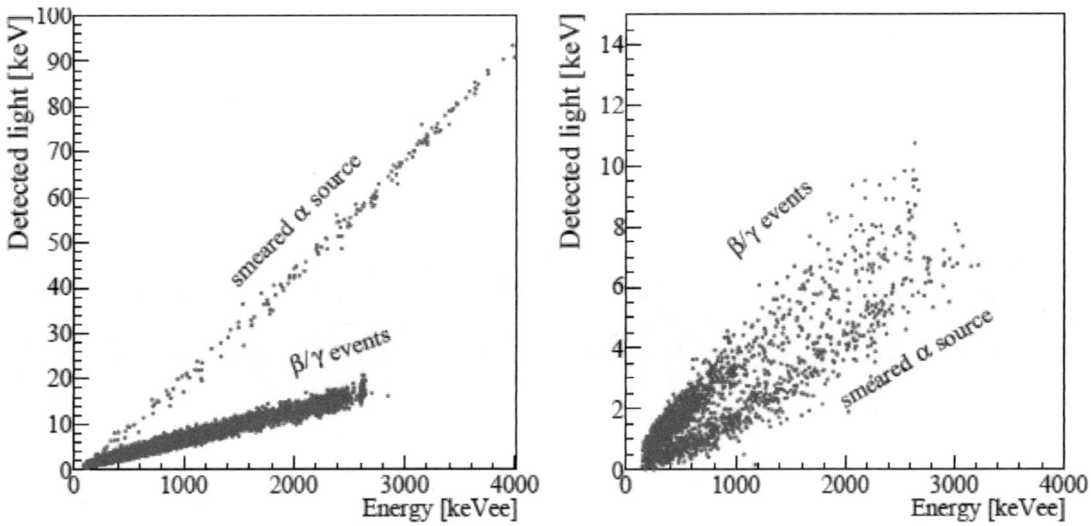

Figure 2. Light vs. Heat scatter plots corresponding to calibration measurements (external ^{232}Th gamma source and smeared ^{238}U alpha source) acquired with a ZnSe #3 before (left) and after (right) thermal treatment. The significant decrease in LY both for alphas and betas/gammas particles and the inversion of the QF for alphas is apparent.

As one can see from figure 2, the LY for alphas and betas/gammas is inverted. The LY for betas/gammas changed from 6.4 ± 0.1 keV/MeV for the "as-grown" ZnSe crystal to a value of 3.2 ± 0.5 keV/MeV for the ZnSe crystal after TT. While the LY for alpha particles decreased more significantly from 29.70 ± 0.17 keV/MeV for the "as-grown" crystal to 2.2 ± 0.2 keV/MeV for the thermally treated ZnSe crystals. It should be stressed, that the LY is changed not only for external alpha particles of calibration source, that reacts with a thin layer of the crystal surfaces, but leads to the significant reduction of the LY for gammas in 2-3 times, which interacts with the entire volume of the crystal. Based on this we can conclude that TT changes the bulk properties of the ZnSe crystals. Such a situation where the characteristic LY values for different type of particles become very close is unacceptable for experiments that rely on the this parameter for particle discrimination.

5. Discussion and summary

The optoelectronic properties of A^2B^6 semiconductors, as e.g. ZnSe, are mainly determined by their defects structure caused by deviation of stoichiometry and the presence of impurities. Despite the huge number of articles dedicated to study the luminescence centers in ZnSe crystals, the interpretation of its luminescence spectrum is still contradictory [18,19,20].

Since all crystals used in these measurements were produced from high purity initial materials (99.9995% grade) we can exclude the presence of significant amount of impurities that can be responsible for the formation of non-radiative recombination centers, defects, and its complexes. But, recalling that the main emission band of ZnSe peaks at 630 nm, the induced light is supposed to be due to the radiative recombination process that occurs within $\{Zn_i\text{-}V_{Zn}\}$ complex [21]. In terms of this model the electron recombines with the hole captured by the V_{Zn} center. The annealing of ZnSe crystal at high temperature destroys such complexes due to the high mobility of Zn atoms diffusing out of the crystal volume. But single defects of V_{zn} are thermally unstable, and in order to compensate charge can forms a thermally more stable and immobile $\{V_{Zn}\text{-}I_{se}\}$ complex. However, this situation looks similar to the annealing of ZnSe crystals in Te or Se vapors, i.e. introduction of isovalent impurities, when quenching of luminescence is interpreted by the formation of recombination center, which cause the strong photo-induced bleaching.

The reduction of ZnSe specific resistivity after thermal treatment indicates appearance of large amount of interstitial Zn, which is the origin of charge carriers, namely electrons. The same effect is observed after annealing of ZnSe crystals in Zn vapors. This fact of the specific resistivity reduction also demonstrates the thermal instability of the $\{Zn_i\text{-}V_{Zn}\}$ defects complex which was destroyed by annealing at 900°C. Based on our observations, we conclude that the $\{Zn_i\text{-}V_{Zn}\}$ defects complex is responsible for the fast radiative recombination process, while formed after thermal treatment defects complex, most probably the $\{V_{Zn}\text{-}I_{se}\}$, demonstrate quenching of luminescence and enlarge kinetics of radiative recombination process. The details on the nature of defects, which are responsible for the luminescence in ZnSe crystals, definitely needs further investigation.

Moreover, it should be mentioned, that scintillating bolometers are suitable tool in order to study fundamental properties of materials at low temperatures. Indeed, this technique is able to evaluate which part of the deposited energy is converted into emitted light, and which is dissipated in the crystal in form of heat.

We have demonstrated that the anomalous QF for alpha particles in ZnSe crystals is due to a certain type of defect in crystal structure and can be adjusted for a various applications via dedicated thermal treatment. Finally, the "as-grown" ZnSe crystals without any kind of thermal treatment are used for the CUPID-0 experiment to avoid a deterioration of their scintillating and bolometric properties.

References

[1] de Marcillac P et al 2003 *Nature* **422** 876
[2] Cozzini C et al 2004 *Phys. Rev. C* **70** 064606
[3] Casali N et al 2014 *J. Phys. G* **41** 075101
[4] Casali N et al 2016 *J. Low Temp. Phys.* **184** 952
[5] Krause E et al 1994 *J. Crystal Growth* **138** 75
[6] Ryzhikov V D and Starzhinsky N G 1988 Sov. *Ukr. Phys. J.* **33** 818
[7] Birks J B 1951 *Proc. Phys. Soc. A* **64** 874
[8] Haller E E et al 1984 *Neutron transmutation doping of semiconducting materials*, ed R D Larrabee (Plenum Press, New York) p 21
[9] Arnaboldi C et al 2002 *IEEE Trans. Nucl. Sci.* **49** 2440
[10] Alessandrello A et al 1998 *Nucl. Instr. Meth. A* **412** 454
[11] Cardani L et al 2013 *J. Instrum.* **8** 10002
[12] Beeman J et al 2013 *J. Instrum.* **8** 05021
[13] Arnaboldi C et al 2004 *Nucl. Instrum. Meth. A* **518** 775
[14] Gatti E and Manfredi P F 1986 *Rivista del Nuovo Cimento* **9** 1
[15] Dafinei I et al 2010 *IEE Trans. Nucl. Sc.* **57** 1
[16] Mikhailik V B et al 2006 *J. Phys. D: Appl. Phys.* **39** 1181
[17] Arnaboldi C et al 2011 *Astropart. Phys.* **34** 344
[18] Yamaguchi M et al 1977 *J. Appl. Phys.* **48** 5237
Caner J et al 1980 *J. Liumin.* **22** 79
Dunstan D J et al 1980 *J. Phys. C* **13** 6409
Jaszczyn-Kopec P et al 1984 *Sol. St. Commun.* **49** 795.
Grimmeiss H G et al 1977 *J. Appl. Phys.* **48** 5122
[19] Baltramiejunas R et al 1985 *Sol. St. Phys.* **27** 371
[20] Baltramiejunas R et al 1993 *Physica* **185** 245
[21] Ryzhikov V et al 2004 *Func. Mater.* **11** 61

IOP Conf. Series: Materials Science and Engineering **169** (2017) 012012 doi:10.1088/1757-899X/169/1/012012

Lithium fluoride thin film detectors for low-energy proton beam diagnostics by photoluminescence of colour centres

R.M. Montereali[1], A. Ampollini[1], L. Picardi[1], C. Ronsivalle[1], F. Bonfigli[1], S. Libera[1], E. Nichelatti[2], M. Piccinini[1], M.A. Vincenti[1]
[1]ENEA C.R. Frascati, Fusion and Technologies for Nuclear Safety and Security Department, Via E. Fermi 45, 00044 Frascati (Rome), Italy
[2]ENEA C.R. Casaccia, Fusion and Technologies for Nuclear Safety and Security Department, Via Anguillarese 301, 00123 S. Maria di Galeria (Rome), Italy
e-mail: rosa.montereali@enea.it

Abstract. Optically transparent LiF thin films thermally evaporated on glass and Si(100) substrates were used for advanced diagnostics of proton beams of energies from 1.4 to 7 MeV produced by a linear accelerator for protontheraphy under development at ENEA C.R. Frascati. The proton irradiation induces the formation of stable colour centres, among them the aggregate F_2 and F_3^+ optically active defects. After exposure of LiF films grown on glass perpendicularly to the proton beams, their accumulated transversal spatial distributions were carefully measured by reading the latent two-dimensional (2-D) fluorescence images stored in the LiF thin layers by local formation of these broad-band visible light-emitting defects with an optical microscope under blue lamp excitation. Taking advantage from the low thickness of LiF thin films and from the linear behaviour of the integrated F_2 and F_3^+ photoluminescence intensities up to the irradiation fluence of ~$5x10^{15}$ p/cm^2, placing a cleaved LiF film grown on Si substrate with the cutted edge perpendicular to the proton beam, the 2-D fluorescence image of the film surface could allow to obtain the depth profile of the energy released by protons, which mainly lose their energy at the end of the path.

1. Introduction

Luminescence properties of point defects in insulating materials are successfully used for solid state light sources and dosimeters. Among them, broad-band light-emitting F_2 and F_3^+ colour centres (CCs) in lithium fluoride, LiF, are well known for their application in tuneable lasers, miniaturised light sources and radiation imaging detectors [1]. Under blue optical pumping in their overlapping absorption bands, their efficient, broad photoluminescence (PL) spans the green-red visible spectral range. Versatile LiF thin-film radiation imaging detectors exploiting the peculiar spectral characteristics of F_2 and F_3^+ defects (two electrons bound to two and three close anion vacancies, respectively) and the radiation sensitivity of the LiF material have been proposed and successfully applied for extreme ultraviolet [2], soft and hard X-ray imaging [1].

The F_2 and F_3^+ laser-active CCs in LiF possess almost overlapping absorption bands peaked around 450 nm, called M band [3]; under light excitation in this spectral range they emit broad PL bands peaked at ≅ 678 nm and ≅ 541 nm, respectively [4], whose intensities can be detected by an optical microscope operating in fluorescence mode. Among the main peculiarities of the luminescent LiF-film solid-state radiation imaging detectors, noteworthy ones are their very high intrinsic spatial resolution, and wide dynamic range. Moreover, due to the stability of F_2 and F_3^+ CCs at room temperature (RT) in LiF, they are quite easy to handle, as insensitive to ambient light.

Recently, their use has been successfully tested for low energy proton beam advanced diagnostics [5,6]. In recent years, the use of hadrons in oncological radiotherapy has seen a remarkable growth because of the excellent ballistic properties of the heavy particles, which lose their energy at the end of the path in tissue, called Bragg peak, with a modest lateral diffusion, preserving the surrounding healthy organs during tumor irradiation. For energies higher than 1 MeV, protons traversing insulating materials lose their energy primarily through ionization and excitation of atoms [7]. Optically transparent LiF thin films thermally evaporated on amorphous and crystalline substrates [8] were used for direct detection of proton beams of energy from 2 to 7 MeV produced by a linear accelerator for protontherapy under development at ENEA C.R. Frascati [9]. The proton irradiation of LiF films induces the local formation of stable CCs, mainly the primary F centres and the aggregate F_2 and F_3^+ ones [5,6].

After exposure of LiF films grown on glass substrates perpendicularly to the proton beams, their accumulated transversal spatial distributions were measured by reading with a conventional optical microscope the latent two-dimensional (2-D) fluorescence images stored in the LiF thin layers by local formation of optically active CCs at irradiation fluence in the range 10^{11}-10^{15} p/cm^2 [5,6].

Placing a cleaved LiF film grown on a Si(100) substrate [8] with the cut edge perpendicular to the proton beam direction, so that the film surface is parallel to the direction of the impinging particles, the 2-D fluorescence images could allow to obtain the depth profile of the energy released by protons, which mainly lose their energy at the end of the path [10,11].

Preliminary experimental results are reported. They show a qualitative agreement with the SRIM (The Stopping and Range of Ions in Matter) software simulations [12], taking into account the polycrystalline nature of the LiF films.

2. Materials and Methods

Exposed samples were polycrystalline LiF thin films, ≅0.8 μm thick, grown by thermal evaporation [8] on glass and Si(100) substrates kept at a constant temperature of 300°C during the deposition process, performed in a vacuum chamber at a pressure below 1 mPa, at ENEA Frascati. The starting material consists of LiF microcrystalline powder (Merck Suprapur, 99.99% pure), heated at about 840°C in a water-cooled tantalum crucible. The evaporation rate, monitored in situ by an INFICON quartz oscillator, was automatically controlled at a fixed value of 1 nm/s during the growth. The substrates were mounted on a rotating sample holder.

Proton beams of 3 and 7 MeV energies were produced by the PL7 model LINAC by ACCSYS-HITACHI [13]. At the output of the machine beamline a 50 μm thick kapton window was placed, which reduced the impinging protons energy to 2.23 and 6.65 MeV, respectively. The average beam current was 1 μA in 60 μs-long pulses at a repetition frequency of 50 Hz. The irradiation fluence covered the range of 10^{11}-10^{15} p/cm^2 by varying the total number of pulses delivered to different LiF samples.

LiF films were irradiated in air at RT in two different mounting geometries: perpendicularly to the proton beam for the detection of the transversal spatial distribution of the proton beam and parallel to the direction of the impinging protons, in order to reveal the Bragg peak position. After exposure, the stored images were read by the fluorescence microscope Nikon Eclipse 80-i C1, equipped with a Hg lamp. The blue emission of the Hg lamp, peaking at 434 nm, was optically filtered in order to simultaneously excite the PL of the F_2 and F_3^+ CCs and an Andor Neo s-CMOS camera was used to acquire the PL images with an 11-bit dynamic range. The Nikon software NIS Elements 4.20 was used to control the optical image acquisition system and the PL image analysis was performed by ImageJ software.

The PL spectra of proton-irradiated LiF films grown on glass were measured at RT by pumping in a continuous wave regime with the 457.9 nm line of an Argon laser, which allows to simultaneously excite the green and red emissions of F_3^+ and F_2 CCs. The PL signal was spectrally filtered by a monochromator and acquired by means of a photomultiplier with lock-in technique. The PL spectra were corrected for the instrumental calibration.

3. Results and Discussion

An important characteristic of the solid state LiF film radiation imaging detectors based on the PL of CCs is the possibility of storing information about the transverse proton beam intensity. Figure 1 shows the PL image of a 7 MeV proton beam stored in a LiF film ≅0.8 μm thick grown on a glass substrate. A quite well collimated proton beam is clearly observed. The formation of stable aggregate F_2 and F_3^+ electronic defects in LiF thin films grown on glass substrates was systematically investigated for low-energy (nominally 3 and 7 MeV) proton beams, in a fluence range from 10^{11} to 10^{15} p/cm^2 [6] in order to study the PL optical response. The integrated CCs PL intensities, as carefully derived from all the fluorescence images, similar to those reported in figure 1, exposed at different irradiation times for both the proton energies, are reported in figure 2. Proton energy loss in the matter continuously increases with depth, reaching the maximum value about at the end of the implantation path, whose position is strongly dependent on the proton energies. Figure 2 shows that the integrated CCs PL signal is proportional to the proton fluence and a linear optical response is found up to a fluence value of 8 x 10^{13} p/cm^2, over which a plateau appears [6]. This behaviour is very similar for 3 and 7 MeV proton beams, although the PL signal is higher for the LiF film exposed at the lower energy at all the investigated fluence values. Such a behaviour can be explained by simulations performed using SRIM software [12]. The inset of figure 2 reports the linear energy transfer (LET) as a function of the proton implantation depth in LiF for 3 and 7 MeV proton beams calculated by SRIM. LET is continuously increasing with depth, reaching the maximum value (Bragg peak) almost at the end of the implantation path, which is higher than the typical thickness of the used LiF films. In LiF films, due to their limited thickness (only 0.8 μm), LET can be considered as constant and only a very small fraction of the total proton energy is lost in them, the rest being deposited in the glass substrate. As shown in the inset of figure 2, the initial LET is lower for 7 MeV protons (12 keV/μm) than for 3 MeV protons (28.4 keV/μm), which is consistent with the lower PL intensity of 7 MeV proton irradiated LiF films.

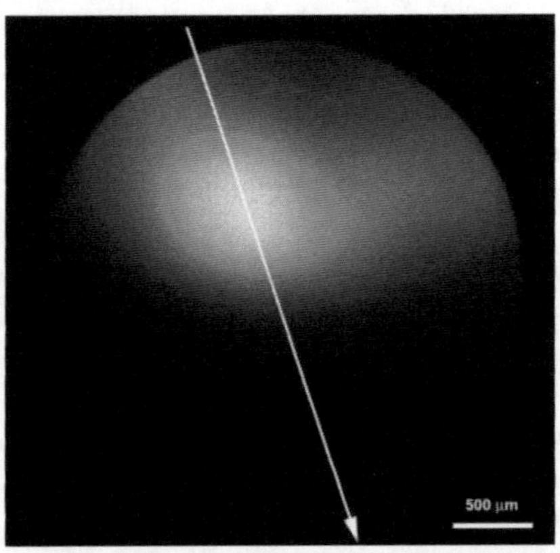

Figure 1 PL optical image of a 7 MeV proton beam transversal section, after passing through a circular aperture, stored by F_2 and F_3^+ colour centres in a LiF film 0.8 μm thick thermally evaporated on a glass substrate.

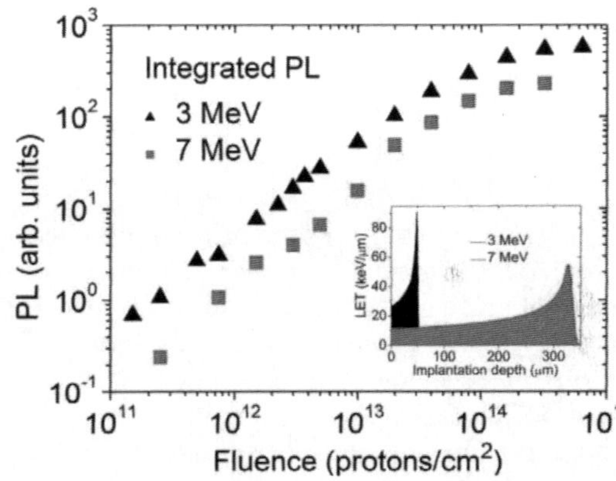

Figure 2. Integrated visible PL signal as a function of the proton fluence in LiF films grown on a glass substrate irradiated by 3 and 7 MeV proton beams as derived from optical images. Inset: SRIM simulation of LET for 3 and 7 MeV proton beams in LiF (see text for details).

In figure 3 the PL intensity profile along the white arrow traced in figure 1 is plotted. Due to the presence of a circular pinhole before the film surface, it is interesting to note that the PL signal value measured outside the pinhole diameter, in the left side of figure 3, corresponding to the "zero" value from a not-exposed film area, is comparable with the signal measured in the not-irradiated area of the exposed LiF film, in the right side of figure 3, outside of the quite well collimated proton beam. This behaviour indicates that the quite good optical quality of the transparent LiF film thermally evaporated on glass does not introduce significant noise contribution in the fluorescence image outside of the proton irradiated area. By comparison, bare glass substrates, irradiated in similar conditions, showed no PL signal in the same measurement conditions, although some darkening was observed in white light. The PL intensities measured by the fluorescence optical microscope were attributed to the local

IOP Conf. Series: Materials Science and Engineering **169** (2017) 012012 doi:10.1088/1757-899X/169/1/012012

formation of stable F_2 and F_3^+ CCs, as confirmed by PL spectra [5,6]. As example, the PL spectrum of a LiF film irradiated by 7 MeV protons at a fluence value of 2.4×10^{12} p/cm^2 is shown in figure 4 as a function of the energy. Its best fit shows that it consists of the superposition of two broad Gaussian emission bands, peaked at around 541 nm and 678 nm, ascribed to F_3^+ and F_2 centres, respectively [3,4]. These emission bands in the PL spectra of proton irradiated LiF films exhibit the same spectral features (peak position and half-width) of irradiated LiF crystals [4] and remained unchanged in the investigated fluence range (not shown).

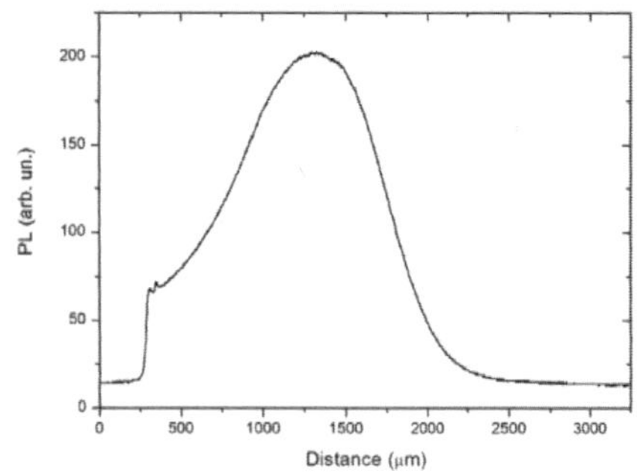

Figure 3. PL intensity profile along the white arrow of figure 1.

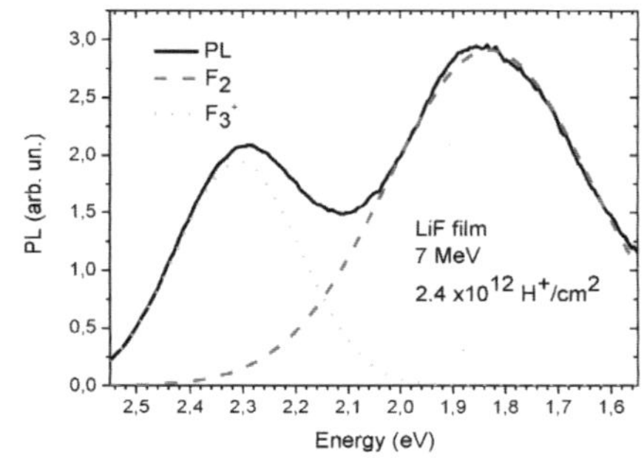

Figure 4. RT laser induced (457.9 nm) PL spectra of a LiF films 0.8 μm thick grown on a glass substrate and irradiated by a 7 MeV proton beam. The F_2 and F_3^+ CCs Gaussian bands obtained from its best fit are also shown.

Figure 5 shows the PL image of a 0.8 μm thick LiF film grown on Si(100) substrate, irradiated by the 7 MeV proton beam at a fluence of 1.6×10^{13} p/cm^2. The spectrally integrated PL emitted by radiation-induced F_2 and F_3^+ CCs, whose local concentrations are proportional to the energy deposited by protons in LiF, exhibits a strong increase at a distance from the cleaved edge of the film, whose amount depends on the proton beam energy. Below saturation, the PL signal is proportional to the local defects concentrations and the strong increase appears to be related to the Bragg peak position. Figure 6 shows the PL intensity profile acquired along the direction perpendicular to the film edge obtained by image analysis of the portion of figure 5 reported below in figure 6.

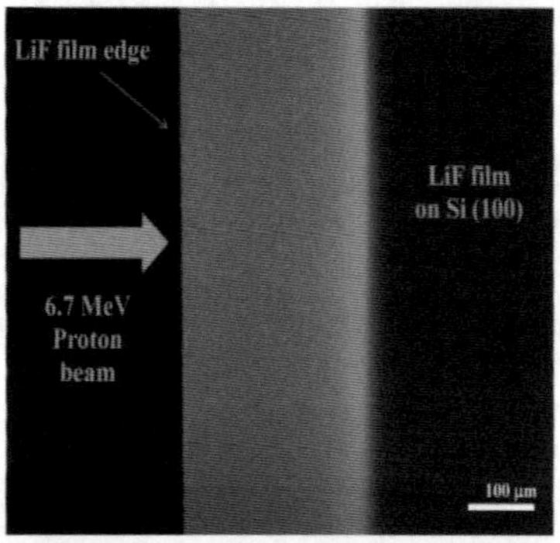

Figure 5. PL image of the top surface of a 0.8 μm thick LiF film grown on a Si(100) substrate, irradiated by a 7 MeV proton beam after cleavage and acquired by the fluorescence microscope.

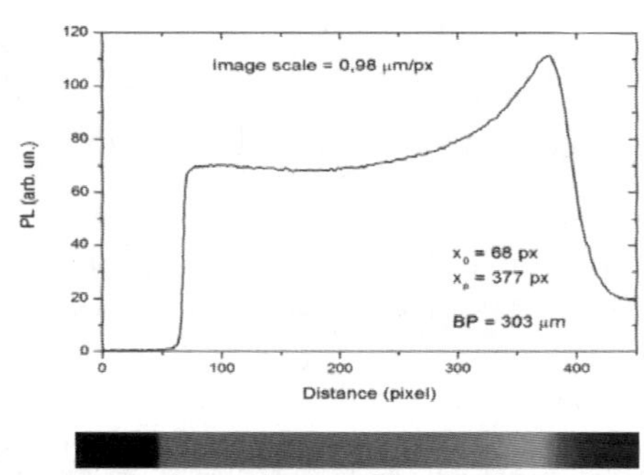

Figure 6. PL intensity profile acquired along the direction perpendicular to the film edge obtained by image analysis of the portion of figure 5 reported in the inset below.

IOP Conf. Series: Materials Science and Engineering **169** (2017) 012012 doi:10.1088/1757-899X/169/1/012012

Radiation induced defects are produced inside a thin layered volume, whose extension is comparable with protons implantation depth, within which they lose all their energy, and a maximum is experimentally found at a distance of ~ 303 μm. This value is comparable with the value found in LiF bulk at 295 μm, according to simulations performed by SRIM [12]. Although the good qualitative agreement, it should be also taken into account that the polycrystalline nature of LiF thin films could imply a reduced density of the LiF films, which can be considered as an aggregate of grains with air interstices [5,8,14].

From an independent estimate, obtained by optical reflectance spectra (not shown), a mean value of 2.609 g/cm^3 was derived. Figure 7 shows Bragg peak position depth values as a function of the proton energy as obtained from the SRIM simulation for a LiF film with the reduced density. The energy corresponding to the distance of 303 μm is too high with respect to the nominal one, whose nominal spread is only +/- 100 keV. On the other hand, it should be highlighted that the Bragg peak depth is extremely sensitive to the material density, as can be inferred from figure 8, where the LET values obtained from SRIM simulation at the effective energy of the impinging protons is shown for several LiF material densities, including bulk, and directly compared with the experimental PL depth profile directly obtained from figure 6.

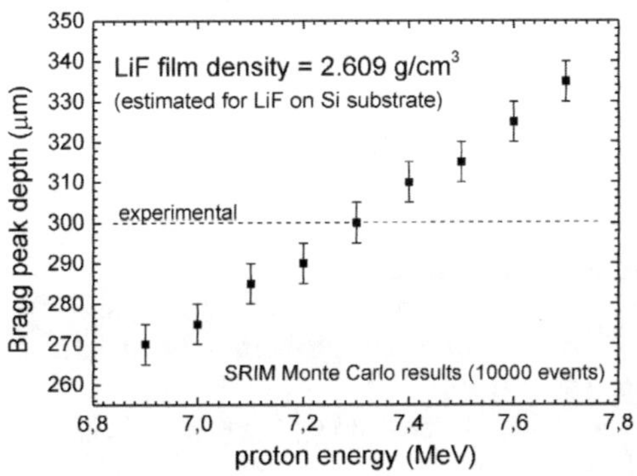

Figure 7. Bragg peak position as a function of the proton energy as obtained from accurate SRIM simulation for a LiF film of density 2.609 g/cm^3.

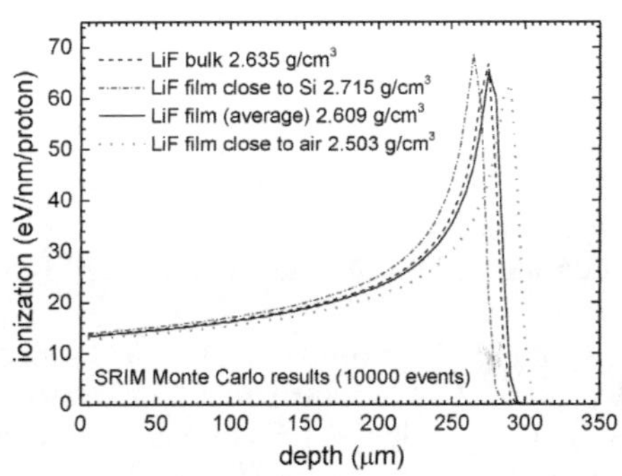

Figure 8. LET values obtained from accurate SRIM simulation for several LiF material densities, including bulk crystals, compared with the experimental PL depth profile in figure 6.

The measured depth at the maximum is higher of all the simulated values and eventually is compatible with a reduced density of 90.5%, which corresponds to 9.5% of voids in the films. However other phenomena, apart the polycrystalline nature of the used LiF samples, can affect the obtained results, among them irregularities in the cleaved edge, as well as PL scattering and image spatial resolution in the optical reading. Moreover, phenomena of diffusion and agglomeration of F CCs are expected to modify the defect profiles with respect to electronic energy loss in LiF crystals [11].

Saturation effects in the CCs PL response as a function of fluence (see figure 2) are responsible for the reduced differences between the PL values measured at the initial depth and on the maximum in the investigated conditions. Other complex phenomena, related to the high proton energy release at the end of the track (nuclear collision) cannot be excluded.

4. Conclusions

Although the PL signal of CCs in proton-irradiated LiF films is at least one order of magnitude lower than in crystals [6] because only a small fraction of the total proton energy is released in it, the rest being deposited in the substrate, optically transparent thin LiF films are able to store information about the proton beam intensity (high sensitivity) with a high spatial resolution (smooth profile) and to reveal even subtle intensity differences (high dynamic range). Despite of their low thickness, at high proton fluencies, 2-D accumulated transversal spatial distributions were carefully measured by reading the latent two-dimensional (2-D) fluorescence images stored in the LiF thin films by local formation of aggregate F_2 and F_3^+ CCs broad-band visible light-emitting defects with an optical microscope under blue lamp excitation. The same reading technique was used to obtain the PL profiles along the proton trajectory, by using cleaved LiF films grown on Si(100). A view of Bragg's profile was experimentally obtained, in qualitative agreement with SRIM simulation and further refinements and experiments are under way.

All the presented results are very encouraging for advanced dose diagnostics of proton beams at high spatial resolution with LiF film detectors based on the visible PL of radiation-induced CCs.

Acknowledgments

Research carried out within the TOP-IMPLART (Oncological Therapy with Protons - Intensity Modulated Proton Linear Accelerator for RadioTherapy) project, funded by Regione Lazio, Italy.

References

[1] Montereali R M, Bonfigli F, Piccinini M, Nichelatti E, Vincenti M A 2016 *J. Lumin.* **170** 761
[2] Baldacchini G et al. 2005 *Rev. Sci. Instrum.* **76** 113104_1
[3] Nahum J and Wiegand D A 1967 *Phys. Rev.* **154** 817
[4] Baldacchini G, De Nicola E, Montereali R M, Scacco A and Kalinov V 2000 *J. Phys. Chem. Solids* **61** 21
[5] Piccinini M, Ambrosini F, Ampollini A, Picardi L, Ronsivalle C, Bonfigli F, Libera S, Nichelatti E, Vincenti M A and Montereali R M 2015 *Appl. Phys. Lett.* **106** 261108
[6] Piccinini M, Ambrosini F, Ampollini A, Carpanese M, Picardi L, Ronsivalle C, Bonfigli F, Libera S, Vincenti M A and Montereali R M 2014 *J. Lumin.* **156** 170
[7] Agullo-Lopez F, Catlow C R A, Townsend P D 1988 *Point Defects in Materials* (London, Academic Press) p 56
[8] Montereali R M 2002 *Handbook of Thin Film Materials* vol 3, ed H S Nalwa (San Diego: Academic Press) chapter 7 pp 399-431
[9] Ronsivalle C et al. 2011 *Eur. Phys. J. Plus* **126** 7 68
[10] Abu-Hassan L H and Townsend P D 1986 *J. Phys. C, Solid State Phys.* **19** 99
[11] Perez A, Balanzat E and Dural J 1990 *Phys. Rev. B* **41** 7 3943
[12] Ziegler J F, Ziegler M D and Biersack J P 2010 *Nucl. Instrum. Meth. B* **268** 1818
[13] Ronsivalle C et al. 2015 *Eur. Phys. Lett.* **111** 14002
[14] Montecchi M, Montereali R M and Nichelatti E 2001 *Thin Solid Films* **396**, 264

Mg co-doping effects on Ce doped $Y_3(Ga,Al)_5O_{12}$ scintillator

[1]**Kei Kamada**[1,2], **Akihiro Yamaji**[3], **Shunsuke Kurosawa**[1], **Yuui Yokota**[1], **Yuji Ohashi**[3] **and Akira Yoshikawa**[1,2,3]
[1]New Industry Creation Hatchery Center, Tohoku University, 6-6-10 Aoba, Sendai 980-8579
[2]C&A corporation, 6-6-10 Aoba, Aramaki, Aoba-ku, Sendai 980-8579, Japan
[3]Institute for Material Reseach, Tohoku University, 2-1-1, Sendai,980-8577 Japan
[4]Institute of Physics AS CR,16253 Prague, Czech Republic
E-mail: kamada@imr.tohoku.ac.jp

Abstract. Nonstoichiometric $Lu_3Al_{5+x}O_{12}$ (x= 0.05, 0.15, 0.35, 0, -0.05, -0.15, -0.35) crystals were grown by the μ-PD method. Luminescence and scintillation properties such as absorption, excitation and emission spectra, light yield and decay time were evaluated. Expected anti-site defect related host emission have been observed in 250-420nm. Emission intensity was increased by increasing the nonstochiometry. The x=-0.35 sample showed the highest light yield of around 12000 photons/MeV and slowest scintillation decay time of 1.96μs.

Introduction

Scintillators coupled with photodetectors were widely used in radiation imaging applications such medical imaging, security, high energy physics, astrophysics, oil well logging, etc. Cerium doped garnet single crystals are promising candidates for scintillator applications. For example, Ce-doped $Lu_3Al_5O_{12}$ (Ce:LuAG) , Ce:$Gd_3Al_2Ga_3O_{12}$ (GAGG) and Ce: $Y_3Al_5O_{12}$ (YAG) single crystal was reported as a prospective scintillator material[1-4]. Ce:YAG was reported with scintillation properties of a scintillation response of about 91.6 ns, and light yield of about 16,200 phot/MeV[3]. Ce:YAG is also used as a phosphor in white LEDs[5,6]. The function of the Ce:YAG phosphor is to absorb the blue light emitted from the blue LED and converts it to yellow light. In multicomponent garnet of Ce:$(Gd,Y)_3(Ga,Al)_5O_{12}$ (GAGG), noticeable light yield over 40,000 phot/MeV at around 500nm with ~80ns decay time was reported. Improvement of scintillation performance was reported due to co-doping with alkali earth AE^{2+} ions in Ce activated scintillators such as, Ce:LSO [7,8], Ce:GAGG[9], Ce:LuAG[10] and Ce:YAG[11]. In the case of Ce:YAG, light yield was improved up to 2,1000 phot/MeV and decay time was accelerated to 31 ns in heavily Mg co-doped samples.
In this report, Mg co-doping effect on scintillation properties in multicomponent garnet of $Y_3(Ga,Al)_5O_{12}$ (YGAG) were investigated. Optical, luminescence and scintillation properties such light yield and decay time were evaluated.

Experimental

[1] Kei Kamada, New Industry Creation Hatchery Center, Tohoku University, 6-6-10 Aoba, Sendai 980-8579

1.1. Crystal growth procedure

A stoichiometric mixture of 4N $MgCO_3$, CeO_2, Y_2O_3, β-Ga_2O_3 and α-Al_2O_3 powders was used as starting material. Nonstoichiometric $(Mg_{0.0002}Ce_{0.05}Y_{0.095})_3(Ga_xAl_{5-x})O_{12}$ (x=1, 2, 3, 4) and Mg non co-doped YGAG single crystals were grown by the micro-pulling-down (μ-PD) method with an RF heating system. The Mg or non co-doped x=1, 2, 3, 4 crystals are named as Mg or non co-doped G1A4, G2A3, G3A2 and G4A1, respectively. A schematic of the μ-PD growth apparatus is given in Refs. [12]. Typical pulling rates were 0.05–0.12 mm/min and the diameter was around 4 mm. Crystals were grown from an Ir crucible under N_2 atmosphere. The seed crystals were <100> oriented YAG crystals. Plates of 4mmϕ x 1mm were cut and polished for the absorption and luminescence spectra measurement.

1.2. Measurements of scintillation properties

Radio-luminescence spectrum at room temperature was measured with the above spectrometer (EI FLS920) excited by 5.5MeV alpha rays from an ^{241}Am source. To determine the light output, we obtained the pulse height spectrum of this crystal under excitation with 662 keV gamma rays from a ^{137}Cs source. The scintillation photons were detected using by a photomultiplier (PMT; Hamamatsu R7600U), and then the signals from the PMT were amplified (ORTEC 113), shaped (ORTEC 572A), and read out with a multi-channel analyzer (AMPTEK 8000A). Additionally, its scintillation decay time excited by the gamma rays (^{137}Cs) was measured with the PMT and an oscilloscope (Tektronix TDS 3052B). The scintillation decay time was calculated from exponential approximation.

Results

1.3. Crystal growth procedure

Mg co-doped and non co-doped Ce:YGAG crystals were grown by the μ-PD method. Example photographs were shown in figure 1. The inner part of grown crystals were transparent with yellow color with 3-4 mm in diameter and 10-40 mm in length. The surface roughness was coming from thermal etching during crystal growth.

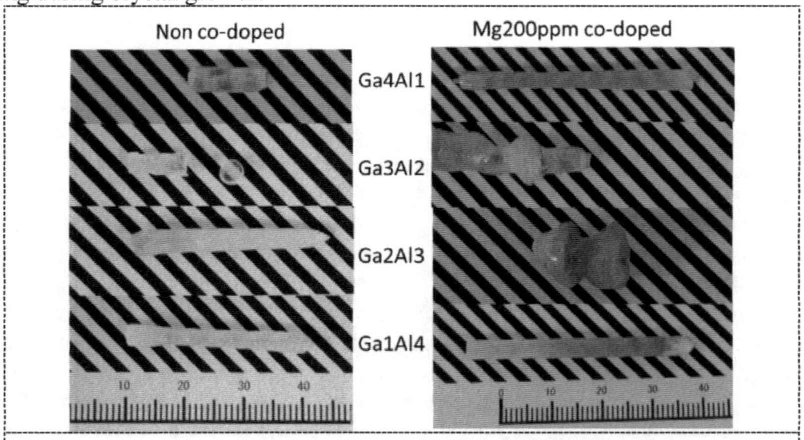

Figure 1. Example photographs of the grown Mg co-doped and non co-doped Ce:YGAG crystals

1.4. Luminescence and gamma-ray response

The absorption spectra of the sample set are presented in Fig. 2(left). In addition of the 4f-5d1,2 absorption bands of Ce^{3+} center at 420 and 350 nm, respectively, the smooth Ce^{4+} CT absorption below 350nm is clearly enhanced by Mg co-doping, similarly to the previous repot in garnet scintilltors [10, 11]. Radioluminescence (RL) spectra of the Mg co-doped and non co-doped G2A3 samples were showed in figure 2(right). There is no energy shift of emission band at 500nm by Mg co-doping.

The typical energy spectra of the Mg co-doped and non co-doped G1A4 samples excited by ^{137}Cs at room temperature are shown in figure 3(left). The Mg co-doped sample showed 180% higher light yield value compared to that of the non co-doped one. The light yield of the sample was calibrated from the ^{55}Fe direct irradiation peak to the APD. Such direct irradiation generates 5.9 keV/3.6 eV = 1640 electron-hole pairs. Based on this value, light yield of Mg co-doped sample was calculated to be approximately 36,000 phot/MeV after correcting the QE, which was 80% at 500 nm. Scintillation decay curves of the Mg co-doped and non co-doped G2A3 samples were shown in figure 3(right). The scintillation decay curves were fitted by second exponential. The decay time was accelerated by Mg co-doping. The Mg co-doped and non co-doped G2A3 samples shows decay time of 19.3ns(80%) and 61.0(20%) respectively. Table 1 shows the scintillation properties. Mg co-doping in G2A3 was most effective for improvement of light yield. According to the previous report, band gap increased and the position of Ce^{3+}4f5d$_1$ level shift to higher energy as decreasing Ga concentration. It is thought that position between conduction band, electron trap coming from anti-site defects, Ce^{3+}4f5d$_1$, and Ce^{4+} levels was optimized at G2A3 composition. The Ce^{4+} center effectively competes with electron traps and contributes to the enhancement of light output and acceleration of decay time.

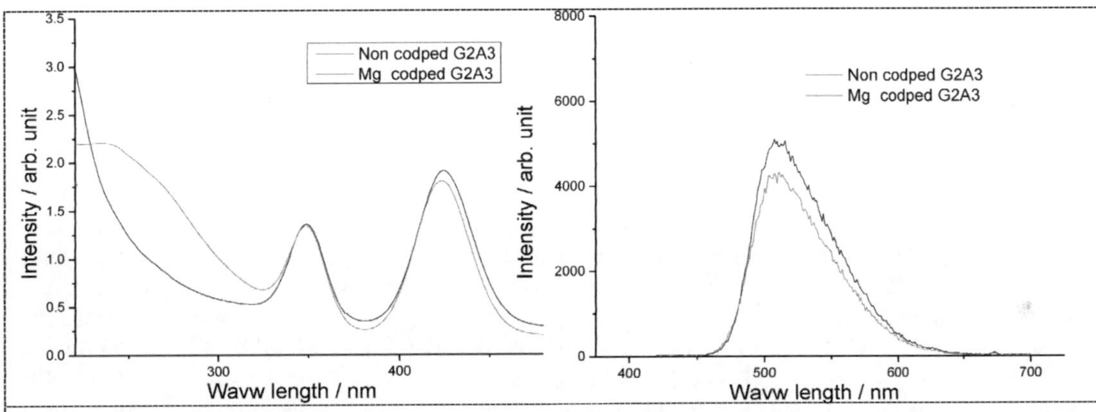

Figure 2. Absorption (left) and radioluminescence spectra (right) of the grown Mg co-doped and non-codoped G2A3 crystals.

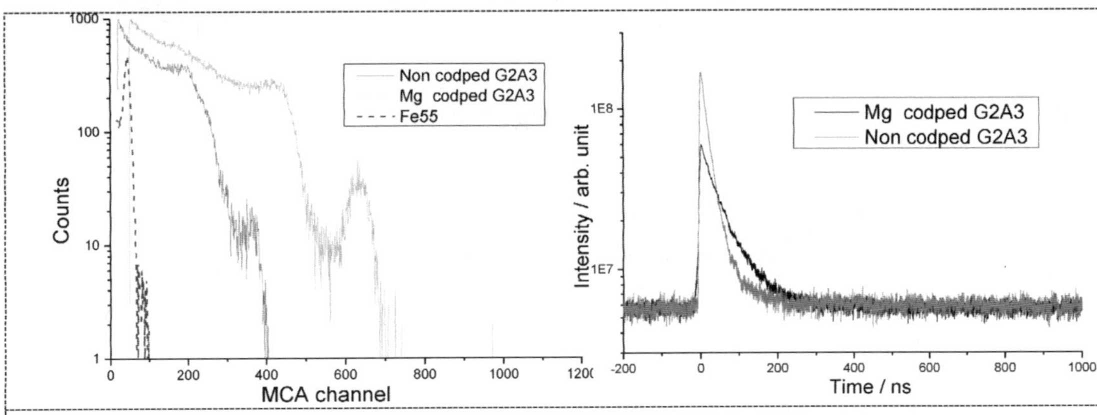

Figure 3. Energy spectra of the grown Mg co-doped and non-codoped G2A3 crystals (left) and decay curves of the grown G2A3crystals (right) excited by a 662 keV gamma ray.

Table 1. Scintillation properties of the grown Mg co-doped and non co-doped YGAG crystals

	Mg concentration (mol.ppm)	Emission wavelength (nm)	Light yield (phot/MeV)	1st decay time (ns)/ratio(%)	2nd decay time (ns)/ratio(%)
G0A5	0	480	16,200	91.6/77	464/23
G1A4	0	489	12,900	7.9/57	15.3/43
G2A3	0	501	13,600	12.7/3	57.3/97
G3A2	0	509	33,400	25.5/10	53.9/90
G4A1	0	517	22,500	38.1/26	79.2/74
G0A5	200	480	19,500	89.3/90	544/10
G1A4	200	492	16,400	19.8/11	51.2/89
G2A3	200	503	36,000	19.3/80	61/20
G3A2	200	510	38,800	11.3/4	54.4/96
G4A1	200	520	29,600	-	68.5/100

Conclusion

The Mg co-doped and non co-doped Ce:YGAG single crystals were grown by the μ-PD method and their optical, luminescence and scintillation characteristics were measured. The absorption of the Ce^{4+} CT band below 250 nm was observed at the Mg co-doped samples. The scintillation decays were accelerated by Mg co-doping and dominant decay time decreased down to 19.3 ns. Maximum light yield value of 38,800 phot/MeV was observed at G3A2 sample. The most significant increase of light yield value due to Mg-codoping was obtained for G2A3 sample, the light yield value of which was 36,000 phot/MeV. These results indicate that Mg co-doped G3A2 and G2A3 can be promising scintillator for application which require high light yield and fast timing resolution without self-radiation such as gamma camera and high energy physics applications. In the near future we will report about the bulk crystal growth of Mg co-doped Ce: YGAG by the Cz method. Improvement on scintillation response is expected due to much higher quality of Czochralski-grown crystal samples.

References

[1]. M. Moszynski, et al, IEEE Trans. Nucl. Sci. 44, (1997) 1052.
[2]. K. Kamada, et al, Crystal Growth & Design, 11 (2011) 4484.
[3]. J.A. Mares, et al.Opt.Mater. 24 (2003) 281.
[4]. M. Nikl, et al, Adv. Opt. Mater. 3, (2015) 463
[5]. K. Bando, et al. J. Light Vis. Environ., 22 (2008) 1.
[6]. J. Meyer, et al, Adv. Opt. Mater. 3, (2015) 424
[7]. C. L. Melcher, et al. IEEE Trans. Nucl. Sci., 39 (1992) 502.
[8]. M. A. Spurrier, et. al, J. Cryst. Growth 310 (2008) 2110.
[9]. K. Kamada, et al, Opt. Mater. 41(2015)63-66
[10]. K Kamada, et al.Nucl. Instrum. Methods Phys. Res. A, 11 (2015)9
[11]. A. Nagura, et al. Japan.J. App. Phy.54 (2015) 04DH1
[12]. Yoshikawa, et al. J. Crystal Growth 205 (1999)305-

Persistent luminescence in powdered and ceramic polycrystalline $Gd_3Al_2Ga_3O_{12}$:Ce

G Dosovitskiy[1], A Fedorov[2], V Mechinsky[3], A Borisevich[3], A Dosovitskiy[4], E Tret'jak[5], M Korjik[2,3]

[1] Institute of Chemical Reagents and High Purity Chemical Substances IREA, 107076 Moscow, Russia
[2] Radiation Instruments and New Components, 220113 Minsk, Belarus
[3] Research Institute for Nuclear Problems, 220030 Minsk, Belarus
[4] NeoChem, 117647 Moscow, Russia
[5] Research Institute for Physical Chemical Problems, 220030 Minsk, Belarus

E-mail: george.dos@gmail.com

Abstract. This paper studies powders of $Gd_3Ga_3Al_2O_{12}$:Ce, a promising scintillator composition, as a possible object for express pre-characterization of scintillation kinetics and level of persistent luminescence. Garnet phase powders with uniform microstructure, consisting of 1-2 µm grains, were obtained by co-precipitation approach. It was shown, that both scintillation decay time and presence of persistent luminescence are influenced by both powder thermal treatment temperature and strong Ga deficit.

1. Introduction

Scintillator crystals and phosphors based on Ce-activated mixed garnets $Gd_3(Ga,Al)_5O_{12}$ (GGAG) get increasing interest recent years [1], as they have promising applications on large markets, such as medical imaging. Light yield of 55000 Ph/MeV was obtained for $Gd_3Ga_3Al_2O_{12}$:Ce transparent ceramic [2] and single crystal scintillators [3]. At the same time, strong persistent luminescence was observed for $Gd_3(Ga,Al)_5O_{12}$:Ce ceramics and single crystals grown from raw materials of 99,995% or better purity [2,4]. Moreover, garnets with Al/Ga solid solution turned out to be promising materials for long glowing phosphors – strong persistent luminescence was observed in $Y_3(Ga,Al)_5O_{12}$:Ce,Cr ceramic samples [5]. High level of afterglow is a major drawback in application of this material in scintillation detectors, so studying this effect and its minimizing is a mandatory and an important issue for its future application.

A number of works was dedicated to optimizing host compound composition. Ogięgło et al [6] found the most intensive photoluminescence in powder samples of $Gd_3Ga_xAl_{5-x}O_{12}$:Ce row at x = 2. For single crystalline materials the most used composition of choice is $Gd_3Ga_3Al_2O_{12}$:Ce [3,7-10].

As a complex oxide solid solution, GGAG system allows wide composition variations. Scintillation kinetics depending on composition was studied in a number of works [3,7-10], results are summarized in Table 1. One can see that kinetics data differ for identical compositions. Inhomogeneous element distribution could have effect on scintillation properties, such inhomogeneity was reported e.g. for complex $(Lu,Gd)_3(Ga,Al)_5O_{12}$:Ce garnet crystals grown by micropulling-down technique [7]. One of possible reasons for that is incontrollable Ga content in a crystal. According to thermodynamic data, maximum evaporation rate of Ga oxide could reach $5*10^{-5}$ – $5*10^{-3}$ mol*cm^{-2}*s^{-1} in a temperature range 1800-2000 °C in oxidizing conditions [11]. It is known, that Gd_2O_3-Ga_2O_3 system has a substantial homogeneity range around a garnet phase – 4% according to [12] and 2,5% according to [13]. Wide homogeneity region accommodating Ga deficit was found for Y_2O_3-Gd_2O_3-Ga_2O_3-Al_2O_3 system in [2]. Certain homogeneity region was found for Gd_2O_3-Ga_2O_3-Al_2O_3 system as well [14], which means that scintillator could accommodate a lot of defects, caused by Ga deficit, and still comprise a single-phase garnet.

Table 1. Scintillation kinetics of GGAG, depending on composition, basing on data from [3,7-10].

Reference:		Composition: $Gd_3Ga_xAl_{(5-x)}O_{12}$:Ce			
		x=3	x=2.7	x=2.4	x=2
[3]	LY, Ph/MeV	55 000	58 000	46 000	
	τ_{sc}, ns (P, %)	97 (80) 353 (20)	172 (88) 1932 (12)	138 (71) 649 (29)	
[7]	LY, Ph/MeV	42 217			45 931
	τ_{sc}, ns (P, %)	52.8 (73) 282 (27)			221 (100)
[8]	LY, Ph/MeV	56 500	55 600	44 600	
	τ_{sc}, ns (P, %)	150(61) 490 (39)	150(61) 490 (39)	130(42) 620(31) 6700(27)	
[9]	LY, Ph/MeV	45 260		40 740	
	τ_{sc}, ns (P, %)	89(73) 286(27)		136(69) 647(31)	
[10] (10x10x10 mm)	LY, Ph/MeV	35 600			
	τ_{sc}, ns (P, %)	156(76) 565(24)			
[10] (2x2x10 mm)	LY, Ph/MeV	34 700			
	τ_{sc}, ns (P, %)	101(65) 319(35)			

Influence of Ga deficit on persistent luminescence is controversial yet. As could be concluded from [2], Ga deficit could be a possible reason for persistent luminescence in $Gd_{1.49}Y_{1.49}Ce_{0.02}Ga_{2.2}Al_{2.8}O_{12}$ ceramics. In [1], on the contrary, composition shift towards excess of Gd leads to increase of light yield and decrease of afterglow in powders with compositions close to $(Gd_{0.988}Ce_{0.012})_3(Al_{0.56}Ga_{0.44})_5O_{12}$.

Existence of correlation between luminescent properties of GGAG polycrystalline ceramic samples and single crystals of corresponding composition was shown in [15], which justifies using polycrystalline samples for preliminary studies of single crystal scintillators.

In order to define an origin of persistent luminescence in $Gd_3Al_2Ga_3O_{12}$:Ce we have examined microstructure properties and scintillation kinetics of powder samples and ceramics, which are synthesized at temperatures up to 1600 °C. Ga volatility is much lower at these conditions, so powder composition could be better determined then for crystals.

2. Experimental

2.1. Samples preparation

Powders were obtained by co-precipitation technique. Nitrate solutions with 1 mole/l concentrations were prepared from Gd oxide (5N), Ga nitrate (4N), Al nitrate (5N), NH_4-Ce nitrate (4N), then mixed in proportion to match desired composition of $Gd_{2.97}Ce_{0.03}Ga_{3-x}Al_2O_{12}$. This joint solution was added to ammonia solution with concentration 2.5 mole/l at a rate 30 ml/min with constant stirring. Then the precipitate slurry was kept boiling for 30 minutes, then filtered, washed with water, dried at 100 °C, crushed, sieved through 100 μm sieve and calcined at 600 °C and then sintered at 1000-1600 °C in air for 2 hours in molten silica or corundum crucibles.

2.2. Samples characterization
Powders microstructure was observed using Jeol JSM 7100 F scanning electron microscope with EDX detector Oxford Instruments X-max^N for local element analysis. Powder X-ray diffractometer GBC Scientific eMMA with Cu K_α radiation was used for phase analysis. Thermogravimetry and differential thermal analysis were performed using Thermo Scientific SDT Q600 analyzer. Photoluminescence spectra were measured using Lumex Fluorat-02-Panorama spectrometer with a fiber lightguide attachment.

2.3. Scintillation measurements
^{22}Na source with gamma emission lines of 511 keV and 1,27 MeV and activity of 1000 Bq was used for scintillation measurements. Decay curves were registered in 1 μs range using start-stop method with ratio of starts to stops 0.02.

3. Results and discussion

3.1. Composition and microstructure
When precipitation was conducted according to the procedure, which works for YAG precipitation [16], using excess of precipitant 20-50%, then 3-7% of Ga remained in mother solution. This corresponds to Ga hydroxide solubility in ammonia water solutions [17]. Boiling precipitate slurry for 30 minutes allows to decrease Ga losses to no more than 0.5%. Filtered and dried precipitate forms dense xerogel translucent pieces, consisting of 10-20 nm particles (Figure 1) with homogeneous elements (Gd, Ga, Al) distribution according to EDX mapping (Figure 2).

Heating the precipitate up to 1200 °C leads to a mass loss of 28%, which corresponds to processes of release of physically and chemically linked water and precipitate decomposition with transformation of Gd, Ga, Al, Ce hydroxides to oxide form. Exothermal peak is found at 880 °C on DTA curve, which, most probably, corresponds to garnet phase formation (Figure 3). In correspondence with this, powders sintered at 1000-1600 °C contain garnet phase (most close to PDF 73-1371) with peaks seemingly narrowing with temperature increase up to sintering temperature of 1400 °C, which indicates increase of their crystallinity with thermal treatment temperature (Figure 4).

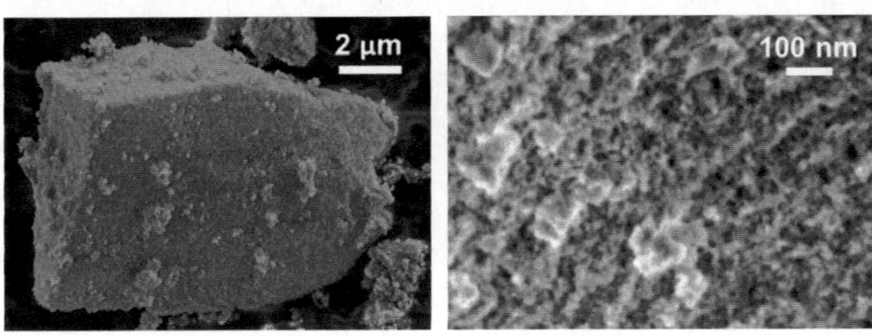

Figure 1. SEM images of a GGAG xerogel.

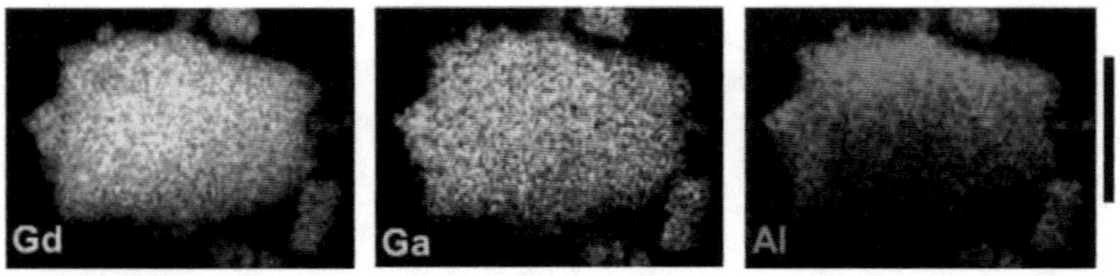

Figure 2. EDX element mapping of a GGAG xerogel particle.

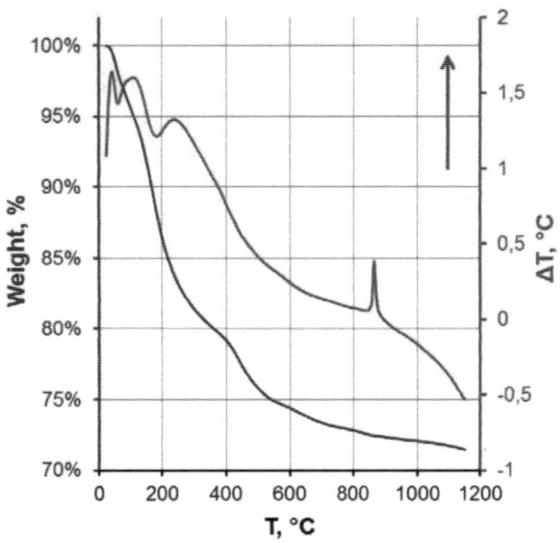

Figure 3. TG-DTA curve for GGAG precipitate, heating rate 10 °C/min.

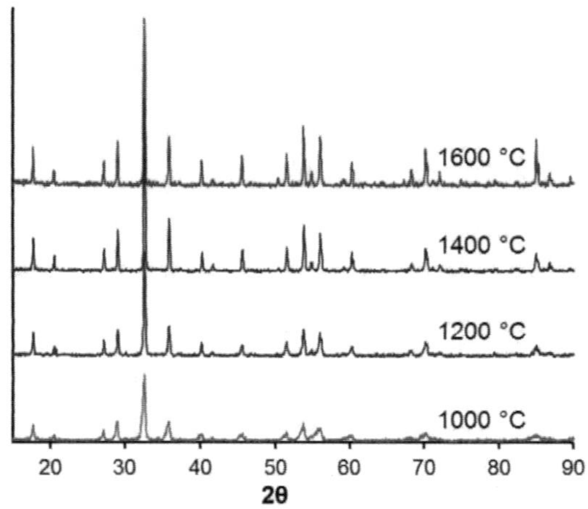

Figure 4. XRD patterns for GGAG powders, sintered at different temperatures

Samples microstructure observed by SEM gradually changes with powder sintering temperature increase (Figure 5). At lower sintering temperatures two types of grains could be seen on images: larger, 0.5-2 µm size and smaller, ~50 nm for 1000 °C and ~100 nm for 1200 °C. Powder particles contain noticeable amount of 50-100 nm pores. Raising sintering temperature leads to a more uniform microstructure: at 1400 °C 0.3-1 µm grains and 0.2-0.3 µm pores, at 1600 °C 1-2 µm grains and up to 1 µm pores (not shown on the image). Faceting of separate grains can be noticed after 1600 °C.

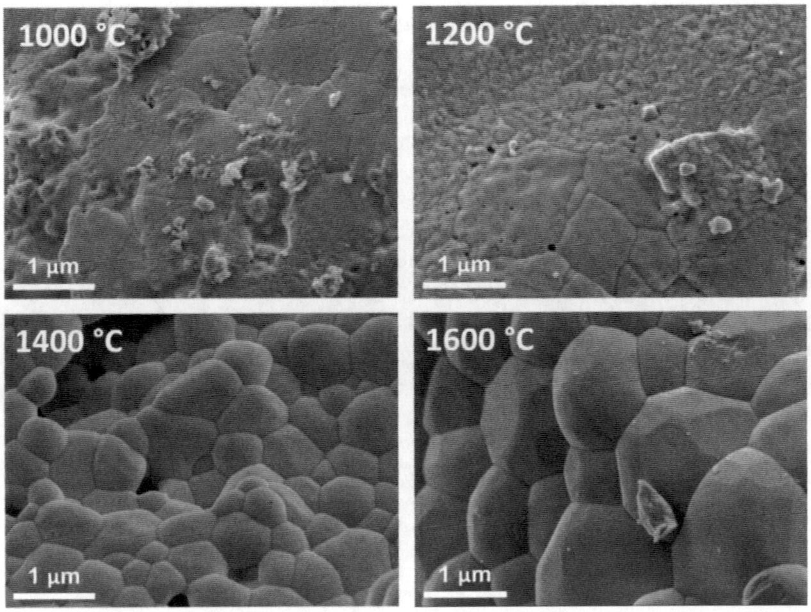

Figure 5. SEM images of a GGAG powder, sintered at different temperatures.

3.2. Luminescent and scintillation properties

Photoluminescence of obtained GGAG powders is typical for Ce-activated garnets with two broad excitation bands (4f-5d transitions $^2F_{5/2} \rightarrow {}^2D_{5/2}$ with maximum at 340 nm, $^2F_{5/2} \rightarrow {}^2D_{3/2}$ with maximum at 445 nm) and one emission band (5d-4f transition $^2D_{3/2} \rightarrow {}^2F_{7/2}$, $^2F_{5/2}$ with maximum at 535 nm) (Figure 6). Ga content variation ±1% does not change bands positions.

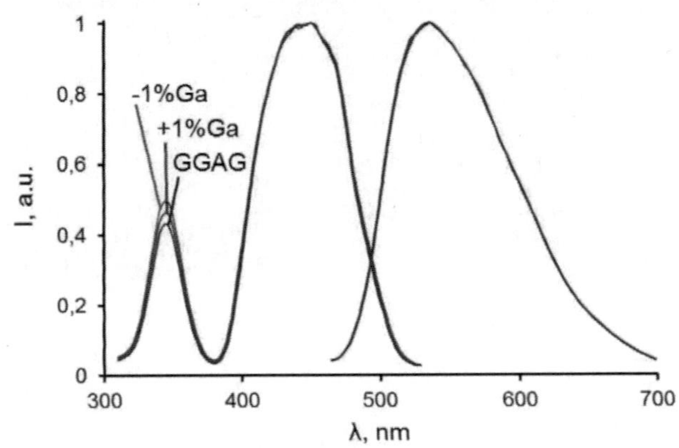

Figure 6. Photoexcitation and photoluminescence spectra (normalized) for GGAG powder with varying Ga content.

Scintillation decay curves were measured to determine samples scintillation kinetics. Background of random coincidences in scintillation kinetics measurements by start-stop method can be used for qualitative estimation of samples persistent luminescence. Light yields were not specially measured, as at the time of the research there was no appropriate technique to adequately estimate light yields on powder samples, to their values are not reported. They could be qualitatively estimated from necessary measurement time to collect enough statistics in kinetic measurements. According to that light yields were the same order of magnitude.

For all samples kinetics contained 2 or 3 exponential components, faster one (τ_1) around 40 ns? slower ones more than 100 ns and averaged scintillation time in the range of 54-81 ns (Table 2).

IOP Conf. Series: Materials Science and Engineering **169** (2017) 012014 doi:10.1088/1757-899X/169/1/012014

Comparing kinetics measurements for samples with different Ga content, one can state, that average scintillation time is higher for samples with 5% Ga deficit (70-80 ns) compared to samples with Ga variation of 1% (55-60 ns). Background level in kinetics measurements is close for them, tenting to be higher for Ga-deficit sample (Figure 7).

Effect of increase of sample thermal treatment temperature can be anticipated from comparison of measurements results for GGAG powder sintered at 1000 °C and GGAG ceramics of the same composition sintered at 1600 °C (Figure 8). It leads to insignificant change of scintillation decay curve and averaged scintillation decay time – 67 ns for 1000 °C powder and 72 ns for 1600 °C ceramics. At the same time, there is also a noticeable difference in background level, indicating that low temperature sample has noticeably lower level of persistent luminescence.

As a proposal for future work, described synthesis approach could be used to prepare homogeneous pre-synthesized raw materials for single crystal growth and test its influence on single crystalline material kinetics and level of persistent luminescence.

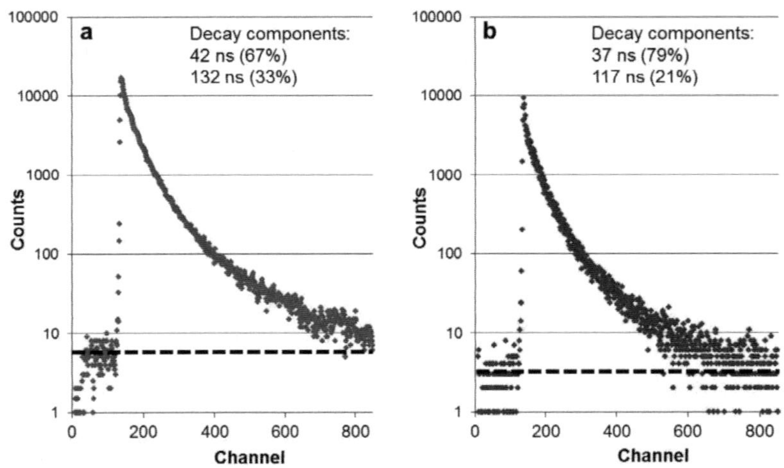

Figure 7. Scintillation decay curves for GGAG powder sintered at 1600 °C with composition $Gd_{2.97}Ce_{0.03}Ga_{2.85}Al_2O_{12}$ (a) and $Gd_{2.97}Ce_{0.03}Ga_3Al_2O_{12}$ (b). 1 channel equals 1 ns. Dashed line indicates background level estimation.

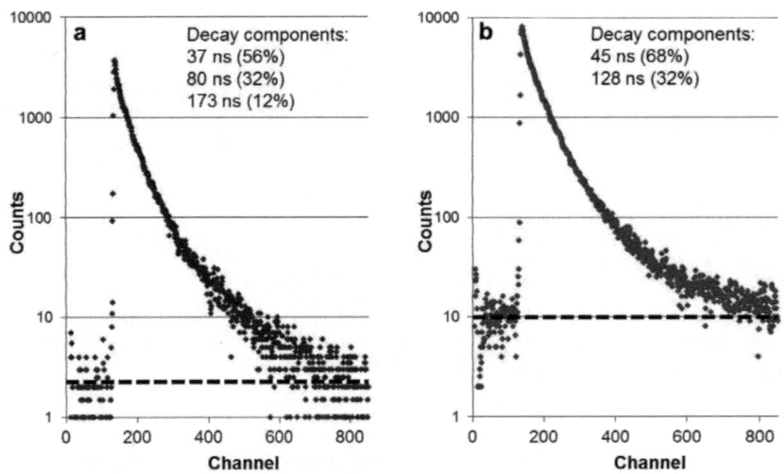

Figure 8. Scintillation decay curves for GGAG powder sintered at 1000 °C (a) and GGAG ceramics of the same composition sintered at 1600 °C (b). 1 channel equals 1 ns. Dashed line indicates background level estimation.

Table 2. Scintillation decay components for GGAG samples of different compositions: powder sintered at 1000 °C, ceramics sintered at 1600 °C, all other – powders sintered at 1600 °C.

Composition:	τ_1, ns	P_1, %	τ_2, ns	P_2, %	τ_3, ns	P_3, %	$\tau_{average}$, ns
$Gd_{2.97}Ce_{0.03}Ga_{3.03}Al_2O_{12}$	48	81%	108	19%			60
$Gd_{2.97}Ce_{0.03}Ga_3Al_2O_{12}$	37	79%	117	21%			54
$Gd_{2.97}Ce_{0.03}Ga_{2.97}Al_2O_{12}$	41	59%	82	41%			58
$Gd_{2.97}Ce_{0.03}Ga_{2.85}Al_2O_{12}$	42	67%	132	33%			70
$Gd_{2.97}Ce_{0.03}Ga_{2.85}Al_2O_{12}$	47	72%	177	28%			81
$Gd_{2.97}Ce_{0.03}Ga_{2.85}Al_2O_{12}$ 1000	37	56%	80	32%	172	12%	67
$Gd_{2.97}Ce_{0.03}Ga_{2.85}Al_2O_{12}$ ceram.	45	67%	128	32%			72

4. Conclusions

Powder polycrystalline materials could be obtained with homogeneous composition, well-defined µm-size grains and well-formed garnet phase. Deficit of Ga in powder samples was shown to slow down scintillation kinetics, increasing averaged scintillation time from 55-60 ns to 70-80 ns. Slightly higher level of background in kinetic measurements was observed in these samples, indicating higher level of persistent luminescence. GGAG ceramics, heat treated at 1600 °C has shown higher level of afterglow, compared to powder heat treated at 1000 °C. So, to use polycrystalline samples as objects for express pre-characterization of single crystal materials, such temperature-dependent parameters as microstructure and crystallinity should be taken into account.

Acknowledgements

Work was supported by Russian Ministry of Science and Education, Subsidy agreement № 14.625.21.0033 dated 27.10.2015, project identifier RFMEFI62515X0033.

References

[1] Kanai T, Satoh M and Miura I 2008 *J. Am. Ceram. Soc.* **91** 456–62
[2] Seeley Z M, Cherepy N J and Payne S A 2013 *J. Cryst. Growth* **379** 79–83
[3] Kamada K *et al.* 2014 *Opt. Mater.* **36** 1942–5
[4] Wu Y, Luo Z, Jiang H, Meng F, Koschan M and Melcher C L 2015 *Nucl. Instr. Meth. A* **780** 45–50
[5] Ueda J, Kuroishi K and Tanabe S 2014 *Appl. Phys. Lett.* **104** 101904 1–4
[6] Ogiegło J M, Katelnikovas A, Zych A, Jüstel T, Meijerink A and Ronda C R 2013 *J. Phys. Chem. A* **117** 2479–84
[7] Kamada K, Endo T, Tsutumi K, Yanagida T, Fujimoto Y, Fukabori A, Yoshikawa A, Pejchal J and Nikl M 2011 Cryst. Growth Des. 11 4484–90
[8] Sibczynski P, Iwanowska-Hanke J, Moszyński M, Swiderski L, Szawłowsk M, Grodzicka M, Szczęśniak T, Kamada K and Yoshikawa A 2015 *Nucl. Instrum. Meth. A* **772** 112–7
[9] Sakthong O, Chewpraditkul W, Kamada K, Yoshikawa A, Szczesniak T, Grodzicka M, Sibczynski P, Moszynski M 2016 *Rad. Meas.* **87** 24–28
[10] Lucchini M T, Babin V, Bohacek P, Gundacker S, Kamada K, Nikl M, Petrosyan A, Yoshikawa A and Auffray E 2016 *Nucl. Instr. Meth. A* **816** 176–83
[11] Lamoreaux R H, Hildenbrand D L and Brewer L 1987 *J. Phys. Chem. Ref. Data* **16** 419–43
[12] Allibert M, Chatillon C, Mareschal J and. Lissalde F 1974 *J. Cryst. Growth* **23** 289–94
[13] Bondar I A, Degtyareva V Ya, Tseitlin P A, Derbeneva T A, Mezentseva L P, Raiskaya L N and Domanskii A I 1988 *Russ. J. Inorg. Chem.* **33** 1820–3
[14] Seeley Z M, Cherepy N J and Payne S A 2014 *J. Mater. Res.* **29** 2332–7
[15] Meng F, Koschan M, Melcher C L and Cohen P 2015 *Mat. Sci. Eng. B* **193** 20–6
[16] Kuznetsova D E, Volkov P A, Dosovitskiy G A, Mikhlin A L, Bogatov K B, Retivov V M and

Dosovitskiy A E 2016 *Russ. Chem. Bull., Int. Ed.* in press
[17] Dymov A M and Savostin A P 1958 *Analitical chemistry of gallium* (M: Nauka, in Rus.) p 14

IOP Conf. Series: Materials Science and Engineering **169** (2017) 012015 doi:10.1088/1757-899X/169/1/012015

An X-Ray Absorption Spectroscopy Study of Ball-Milled Lithium Tantalate and Lithium Titanate Nanocrystals

A.V. Chadwick[1], D.M. Pickup[1], S. Ramos[1], G. Cibin[2], N. Tapia-Ruiz[3], S. Breuer[4], D. Wohlmuth[4] and M. Wilkening[4]

1 School of Physical Sciences, University of Kent, Canterbury, Kent CT2 7NH

2 Diamond Light Source, Diamond House, Harwell Campus, Didcot, Oxfordshire, OX11 0DE

3 Department of Materials, University of Oxford, Parks Road, Oxford OX1 3PH

4 Institute for Chemistry and Technology of Materials, Christian Doppler Laboratory for Lithium Batteries, Graz University of Technology, Stremayrgasse 9, 8010 Graz, Austria

E-mail: a.v.chadwick@kent.ac.uk

Abstract. Previous work has shown that nanocrystalline samples of lithium tantalate and titanate prepared by high-energy milling show unusually high lithium ion conductivity. Here, we report an X-ray absorption spectroscopy (XAS) study at the Ti K-edge and the Ta L_3 edge of samples that have been milled for various lengths of time. For both systems the results show that milling creates amorphous material whose quantity increases with the milling time. The more extensive data for the tantalate shows that milling for only 30 minutes generates ~25% amorphous content in the sample. The content rises to ~60% after 16 hours. It is suggested that it is the motion of the lithium ions through the amorphous content that provides the mechanism for the high ionic conductivity.

1. Introduction

There are numerous reports in the literature that atomic migration in nanocrystals is unusually fast compared to the parent bulk material. In ionic materials the origin for enhanced diffusion has been assigned to atomic disorder along interfaces and surfaces [1-4], however, in many cases, the experimental data are ambiguous. The data for the binary and ternary oxides is particularly varied and is very dependent on the method of sample preparation [5-9] as this controls the microstructure of the material. A simple and rapid method of producing relatively large quantities of nanomaterials is high-energy ball-milling [10-13]. However, it is now clear that for simple binary oxides, like alumina, Al_2O_3 [14] and zirconia, ZrO_2 [15] it produces samples that have a large (tens of per cent) amorphous content; the nanocrystalline core is surrounded by a shell of amorphous material. The effect on the atomic migration of this microstructure is not clear. In contrast, ball-milled ternary oxides can show unusually enhanced ionic mobility.

The first ternary ball-milled system that was studied in any depth was lithium niobate, $LiNbO_3$, and the samples exhibited very high lithium ion diffusivity and conductivity [16-18];

the material in bulk crystalline form is normally regarded as an insulator. ^{7}Li nuclear magnetic resonance (NMR) studies of ball-milled $LiNbO_3$ with particle sizes down to 16 nm showed lithium ion mobility was increased by several orders of magnitude [16]. Nb K-edge X-ray absorption spectroscopy (XAS) of ball-milled $LiNbO_3$ showed a highly attenuated Extended X-ray Absorption Fine Structure (EXAFS) indicative of a significant amorphous content [17]. A very thorough study by Heitjans and co-workers [18] of ball-milled and sol-gel prepared $LiNbO_3$ using a wide range of techniques confirmed the amorphous nature of the surface of the nanocrystals and that this was the origin of the high lithium ion mobility.

Currently there is intense interest in improving the performance of lithium-ion batteries and a necessary requirement of components is high lithium ion mobility [19]. Hence the observations for ball-milled $LiNbO_3$ have led to the study of other similar materials. Ball-milled $LiTaO_3$ with a particle size of about 20 nm exhibits an ionic conductivity five orders of magnitude higher than bulk crystals [20]. Similarly ball-milling Li_2TiO_3 [21, 22] increases the lithium ion conductivity by some three orders of magnitude. As in the case of $LiNbO_3$ the increased lithium ion diffusion is assigned to the generation of amorphous material. In this contribution we confirm and quantify the presence of amorphous material in these two materials using XAS measurements.

2. Methodology

2.1 Materials

Micron sized commercial powders of Li_2TiO_3 and $LiTaO_3$ were ball-milled using the procedures described in previous papers [20-22]. For Li_2TiO_3 only two samples were studied; micron powder and after 8 hours milling. In the case of $LiTaO_3$ micron powder and samples after 0.5, 1, 4, 8 and 16 hours of milling were investigated. For the XAS measurements the powders were mixed with cellulose diluent and pressed into 13 mm diameter pellets.

2.2 XAS measurements

XAS scans were collected for the appropriate edge (Ti K-edge and Ta L_3-edge) at room temperature on beam line B18 at the Diamond Light Source [23]. Data collection used transmission mode with ion chamber detectors. Continuous scanning (QEXAFS) was employed; an individual scan required 180 s and several scans were performed to improve the signal-to-noise ratio. The synchrotron energy and current were 3 GeV and 300 mA, respectively. The beam size at the sample

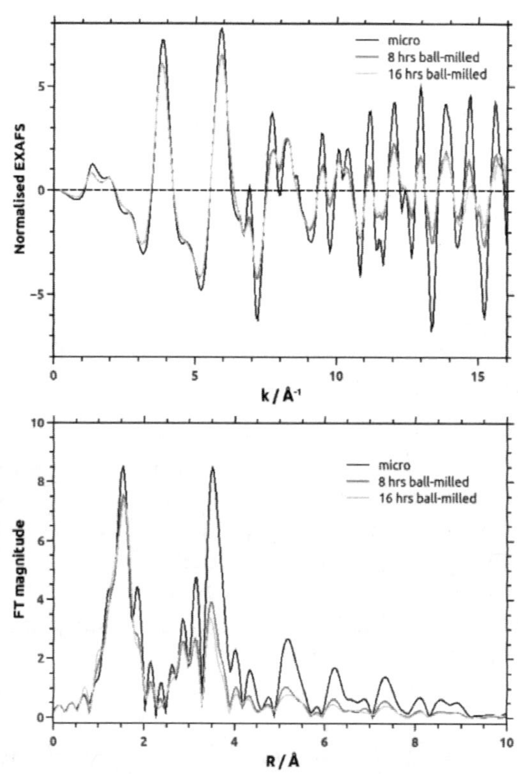

Figure 1: The Ta L_3-edge EXAFS of $LiTaO_3$. Upper – normalised EXAFS. Lower – Fourier transform

was 700 × 700 microns. Powdered samples were mixed with cellulose as a diluent and pressed into 13 mm diameter pellets. The spectra were normalized in Athena and fitted to scattering models in R-space produced by FEFF in Artemis [24].

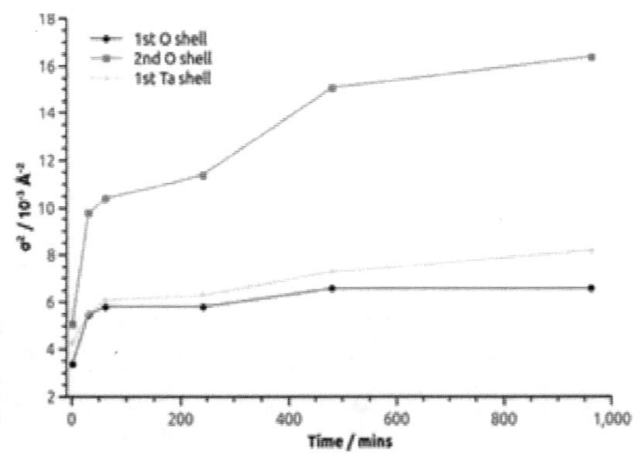

Figure 2: The Ta L_3-edge EXAFS Debye-Waller factors for $LiTaO_3$ as a function of milling time.

3 Results and discussion

3.1 $LiTaO_3$ XAS measurements

All the spectra were analysed to yield the details of the local structure, i.e. bond lengths, *R*, Debye-Waller factors, σ, etc.; the qualitative information is, however, best represented by the Fourier transform (FT) of the k^3 weighted normalised EXAFS and these will be mostly used in this short paper. The data for three samples are shown in Figure 1. There is an attenuation of the EXAFS at high *k* leading to a reduction in the peaks in the corresponding Fourier transform, particularly the Ta-Ta correlation at 3.8 Å. We have analysed the effect in two ways; as a change in the Debye-Waller factor due to disorder and as the effect of an amorphous content. A full analysis with FEFF allowing all the key parameters to float gave very good fits to the spectra. The parameters that changed significantly with milling time were the Debye-Waller factors, σ, for the Ta-O, Ta-O and Ta-Ta correlations at 1.9, 2.0 and 3.8 Å, respectively. These are plotted in Figure 2 in the form of σ^2 versus milling time. It can be seen that for all the correlations σ^2 almost doubles after only 30 minutes milling. This suggests that even this relatively brief milling causes a distortion of the local structure around the Ta ions. After 30 minutes the Ta-O Debye-Waller factors show only a gradual change whereas the Ta-Ta correlation, which is significantly larger, continues to increase steeply with milling time. An explanation of these effects is that milling after 30 minutes causes little changes to the local Ta-O octahedra but the relative arrangement of the octahedra continues to change. This would be expected given that the strong Coulomb interaction between Ta^{5+} and O^{2-} will maintain a relatively rigid local structure. This explanation is also consistent with the shape of the ^7Li NMR quadrupolar powder patterns observed for a sample milled for 30 min, see ref. [20]. After 30 min of milling most of the local electric field gradients in the direct neighbourhood of the Li nuclei are still comparable to those in unmilled, coarse-grained $LiTaO_3$ [20].

An alternative, but not contradictory, approach is to assume the attenuation of the EXAFS is due to the ball-milling generating amorphous material, as in the analysis of the EXAFS of ZrO_2 [15]. This would be mainly reflected in the peak height of the Ta-Ta shell in the Fourier transform. Assuming that the commercial powder prior to milling is 100% crystalline

Table 1

Milling Time/Hour	Estimated Amorphous Content	
	$LiTaO_3$	Li_2TiO_3
0	0%	0%
0.5	~ 25%	
1	~ 41%	
4	~ 44%	
8	~ 55%	~25%
16	~ 60%	

then the height of the Ta-Ta peak in the Fourier transform is representative of a sample with no amorphous content. Hence the relative height of this peak in the milled samples can be used to estimate the degree of crystallinity, i.e. the amorphous content. The results of this rather crude, but useful, analysis are shown in Table 1. This analysis also reveals that milling for as little as 30 minutes drastically affects the sample and generated ~25 % amorphous material.

3.2 Li_2TiO_3 XAS measurements

We have fewer data for this system but they follow the same pattern as those for $LiTaO_3$. The EXAFS results are shown in Figure 3. Again, ball-milling attenuates the EXAFS and there is a decrease in the height of the Ti-Ti peak in the Fourier transform. The same analysis used for $LiTaO_3$ suggests the ball-milled sample contains 25% amorphous material.

It would be unwise to assume that the apparent quantitative differences between the titanate and tantalate after 8 hours ball-milling are meaningful. Other factors, such as the relative hardness of the tow materials, would need to be considered.

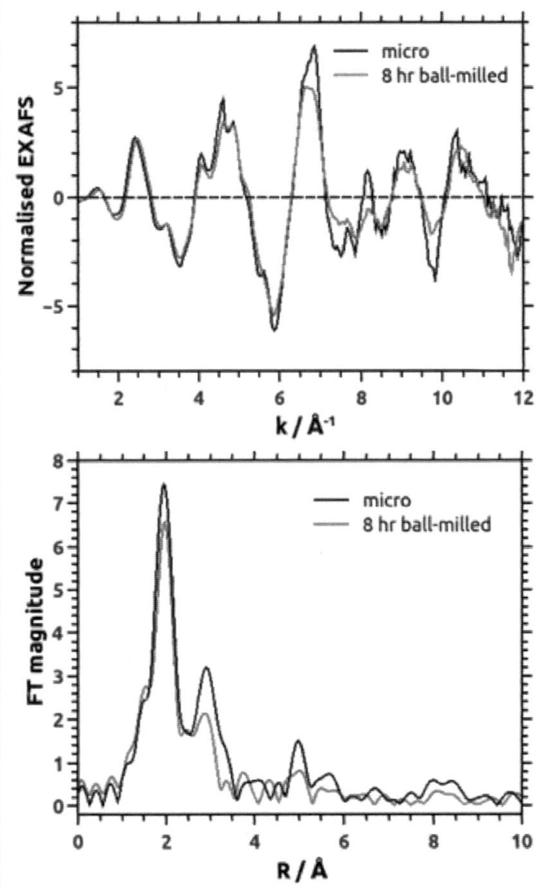

Figure 3: The Ti K-edge EXAFS of Li_2TiO_3. Upper – normalised EXAFS. Lower – Fourier transform

4. Conclusions

There are clear similarities between the current XAS results for $LiTaO_3$ and Li_2TiO_3 and the data for $LiNbO_3$ reported in earlier papers [17, 18]. Hence, a reasonable *general* explanation of the results for these two systems is that ball-milling is creating nanoparticles that have a crystalline core with an amorphous shell.

The more extensive study of $LiTaO_3$ is particularly interesting as it shows that ball-milling for only 30 minutes dramatically attenuates the EXAFS spectrum, indicating a highly disordered sample. This parallels the conductivity measurements of a sample ball-milled for 30 minutes where the magnitude is four orders higher than that for an un-milled sample [20]. However, the conductivity studies revealed some subtle dependence of the microstructure on the time of ball-milling. The longer the milling time the more difficult it is to recover crystallinity by thermal annealing.

Finally, it is interesting to compare the results for the current systems and other oxides with similar data for binary fluorides. XAS studies of ball-milled binary fluorides, like CaF_2

and BaF_2 show virtually no attenuation of the EXAFS spectra, suggesting a very small, if any, amorphous content [25]. The difference is presumably a reflection of the greater strength of metal-fluorine bonds compared to metal-oxygen bonds.

Acknowledgements

We thank the Diamond Light Source for the award of beam time as part of the Energy Materials Block Allocation Group SP14239.

References

[1] Tuller H 2000 *Solid State Ionics* 131 143

[2] Indris S, Heitjans P, Roman H E and Bunde A 2000 *Phys. Rev.Lett.* 84 2889

[3] Guo X, Tan C and Yuan R-Z 1995 *J. Eur. Ceram. Soc.* 15 25

[4] Heitjans P and Indris S 2003 *J. Phys.: Condens. Matter* 15 1257

[5] Chadwick A V and Savin S L P 2009 *Solid State Electrochemistry I*, ed. V V Kharton (Weinheim: VCH-Wiley) p 79

[6] Chadwick A V and Savin S L P 2010 *Low Dimensional Solids*, ed. D W Bruce, D O'Hare and R I Walton (Chichester: Weinheim: Wiley) p 1

[7] Aydin H, Korte C and Janek J 2013 *Sci. Technol. Adv. Mater.* 14 035007

[8] Gerstl M, Frömling T, Schintlmeister A, Hutter H and Fleig J 2011 *Solid State Ionics* 184 23

[9] Schichtel N; Korte C, Hesse, D, Zakharov N, Butz B, Gerthsen D and Janek J 2010 *Phys. Chem. Chem. Phys.* 12 14596

[10] Fecht H J 1995 *Nanostruct. Mater.* 6 33

[11] Koch C C 1997 *Nanostruct. Mater.* 9 13

[12] Cukrov L M, Tsuzuki T and McCormick P G 2001 *Scr. Mater.* 44 1787

[13] Indris S, Bork D and Heitjans P 2000 *J. Mater. Synth. Process.* 8 245

[14] Scholz G, Stosser R, Klein J, Silly G, Buzaré J Y, Laligant Y and Ziemer B, 2002 *J. Phys.: Condens. Matter* 14 2101

[15] Chadwick A V, Pooley M J, Rammutla K E, Savin S L P and Rougier A 2003 *J. Phys.: Condens. Matter* 15 431

[16] Bork D and Heitjans P, 2001 *J. Phys. Chem. B* 105 9162

[17] Heitjans P, Masoud M, Feldhoff A, Wilkening M, 2006 *Faraday Discuss.* 134 67

[18] Pooley M J and Chadwick A V, 2003 *Rad. Eff. Def. Solids* 158 197

[19] Tarascon J-M and Armand M, 2001 *Nature* 414 359

[20] Wilkening M, Epp V, Feldhoff A and Heitjans P, 2008 *J. Phys. Chem. C* 112 9291

[21] Ruprecht B, Wilkening M, Uecker R and Heitjans P, 2012 *PCCP* 14 11974

[22[Brandstätter H, Wohlmuth D, Bottke P, Pregartner V and Wilkening M, 2015 *Z. Phys. Chem.* 229 1363

[23] Dent A J, *et al* 2013 Journal *of Physics: Conference Series* 430 0120231

[24] Ravel B and Newville M 2005 *J. Synchrot. Radiat.* 12 537

[25] Chadwick A V, Düvel A, Heitjans P, Pickup D M, Ramos S, Sayle D C and Sayle T X T, 2015 *IOP Conf. Series-Mat. Sci. and Eng.* 80 #012005

The effect of optical bleaching at selected photon energies on the thermoluminescence of LiF:Mg,Ti

G. Reshes[1], S. Druzhyna[2], S. Biderman[3], I. Eliyahu[4], L.Oster[1], and Y.S.Horowitz[5*]

[1] Physics Unit, Sami Shamoon College of Engineering, Beersheva, Israel
[2] Nuclear Engineering Unit, Ben Gurion University of the Negev, Beersheva, Israel
[3] Nuclear Research Center- Negev, Beersheva, Israel
[4] Nuclear Physics and Engineering Department , Soreq Nuclear Research Center, Yavne, Israel
[5] Physics Department, Ben Gurion University of the Negev, Beersheva, Israel

*Corresponding author. Tel: +972 8 6275294; fax: +972 8 6275294, E-mail address: yigalho@gmail.com (Y.S. Horowitz).

Keywords: Optical absorption, Thermoluminescence, photon bleaching, TLD-100, composite glow peak 5

Abstract. A new interpretation of the optical absorption energy spectrum of LiF:Mg,Ti (TLD-100) is presented in which the 4.0 eV band is composed of two-sub-bands at 3.8 eV and 4.3 eV. The new interpretation is of central importance in the understanding of the geminate recombination mechanisms giving rise to composite glow peak 5 in the thermoluminescence (TL) of TLD-100. Photon bleaches at energies of 3.6 eV and 4.3 eV on samples of gamma irradiated TLD-100 were carried out in order to investigate the correlation of the 4.3 eV absorption band with the electron-hole configuration giving rise to glow peak 5a and the 3.8 eV absorption band with the e-only configuration giving rise to glow peak 5. The 4.3 eV bleach decreases peak 5a dramatically and increases peak 4 (a hole-only configuration). No such effect is observed following the 3.6 eV bleach. The results strongly support the suggested structures giving rise to the 4.3 eV and 3.8 eV bands observed in the optical absorption energy spectrum.

1. Introduction

Recent investigations [1] have suggested that the 4.0 eV optical absorption (OA) band in LiF:Mg,Ti (TLD-100) is actually a composite structure consisting of two bands at 3.84 eV and 4.3 eV. In this new interpretation, the 3.84 eV and 4.3 eV bands replace the broad 4.0 eV band and the 4.45 eV band. The poor fit to the experimental spectrum at ~ 4.45 eV was the reason that prompted various authors [2-3] to include an additional band of unknown origin at this energy as shown in Figure 1. A typical OA spectrum with the new interpretation is shown in Figure 2.

The 5.45 eV band is believed to arise from a trapping center (TC) which acts as the competitor to the 4 eV TC when the sample is heated to produce the TL of composite peak 5. The 5.08 eV band arises from F center absorption and the 4.77 eV band may be F center –related but its exact structure is unknown. The 3.3 eV band is believed to be arise from Mg based dipoles and is correlated with glow peak 2. Both the 3.3 eV band and the 4 eV band are absent in the OA spectrum of 4N single crystal LiF (Figure 3) which establishes their association with the Mg and Ti dopants in TLD-100. The extremely low relative intensity of both glow peaks 5a and 5 in the 4N sample, approximately 0.1% per mgm compared to TLD-100 which lends additional credence to the reasoning that the 4.3 eV entity observed in 4N material do not origin from the same impurities.

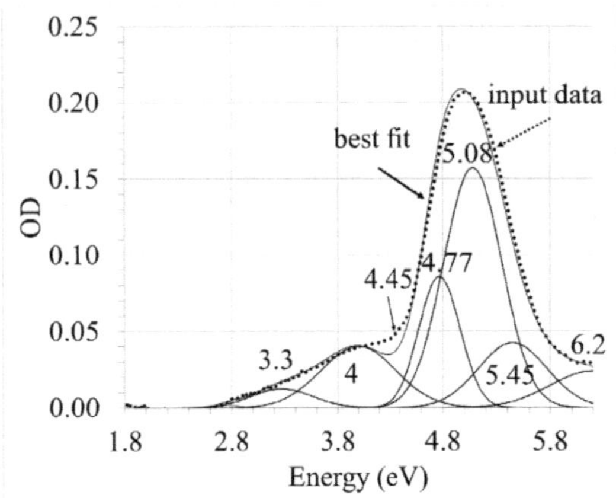

Figure 1. OA spectrum with conventional analysis – 4.0 eV and 4.45 eV OA bands of sample with 0.4mm sample, irradiated to 2500 Gy. Input data (dotted line), fit (solid line). Reproduced from [1].

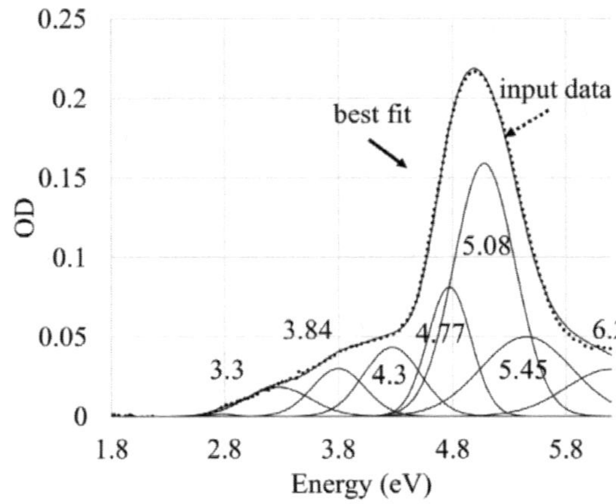

Figure 2. OA spectrum with new analysis – 3.84 eV and 4.3 eV bands of 0.4mm samples, irradiated to 2500 Gy. Input data (dotted line), fit (solid line). Reproduced from [1].

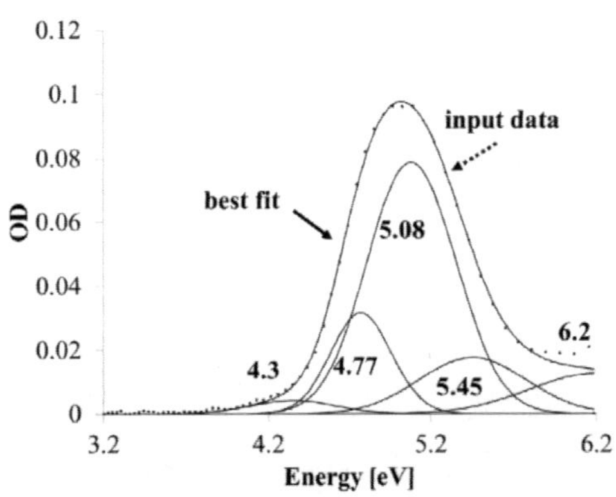

Figure 3. OA spectrum of 0.9 mm 4N single crystal LiF irradiated to 1000Gy.

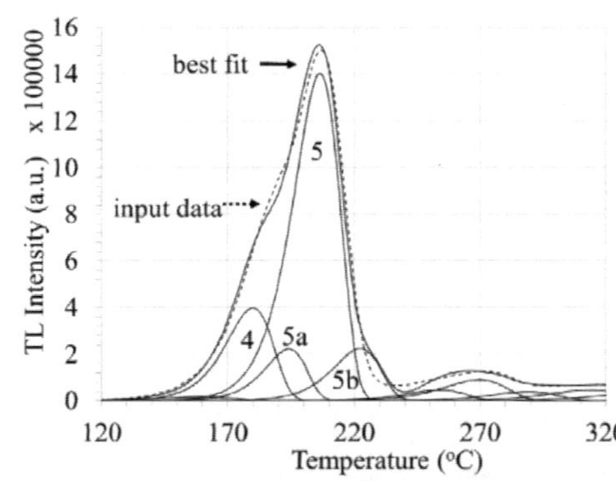

Figure 4. Typical TL glow curve TLD-100 following irradiation and PIA.

The 4.0 eV band has been associated with a trimer trapping center (TC) based on Mg-Livac dipoles and coupled to Ti-OH molecules [2,3]. The 4.45 eV band was considered controvertial in nature and could not be associated with any specific defect [4]. The 4.0 eV OA band has been associated with glow peak 5 in the glow curve of LiF:Mg,Ti (TLD-100) which is also believed to be a composite structure consisting of glow peak 5a (arising from an electron-hole configuration)) and the major peak 5 (arising from an e-only configuration in a spatially correlated trapping center/luminescent center (TC/LC) [5]. It is also believed that the hole-only configuration gives rise to peak 4 in the thermoluminescence (TL) glow curve [6,7]. This TC/LC model has been used to explain many of the ionization density dependent characteristics of these glow peaks [8] but distinct OA bands supporting the interpretation of glow peaks 5a and 5 have not been observed prior to this work. A typical glow curve of TLD-100, following gamma irradiation to 1 Gy and post irradiation anneal (PIA) to 160°C for 6 s, is shown in Figure 4.

2. Materials and Methods

Four samples of TLD-100 (3 x 3 x 0.89 mm^3) were irradiated to 1Gy with ^{137}Cs gamma rays. The samples were pre-irradiation annealed at 400°C for 1 hr followed by slow cooling (50 °C hr^{-1}) to room temperature. Glow curve readout was carried out at a heating rate of 1°C s^{-1} following a post-irradiation anneal (PIA) of 160°C for 6 s which removes the low temperature glow peaks and simplifies the analysis of the effects of the photon bleach on the glow curve. Glow curve deconvolution was carried out with first order peak shapes based on anciliary thermal and optical excitation which isolate peaks 4 and 5 and allow estimates of the peak parameters relatively free of deconvolution uncertainties [9]. The starting parameters of the deconvolution analysis are shown in **Table 1**. The 160°C PIA removes peaks 2 and 3 from the glow curve.

Table 1. Glow peak starting parameters used in the deconvolution analysis.

Peak	Peak Temperature [°C]	Activation Energy [eV]
2	119	1.1
3	162	1.15
4	182	1.9
5a	196	2.3
5	208	2.15
5b	224	2.25
6	241	2.1
7a	256	2
7	271.5	2
8	291	2
9	315.5	2
10	338	1.8
11	366.5	1.8

For the optical bleach experiments the light source employed was a 300 W xenon lamp attached to an Oriel 1/8 m monochromator with 1200 lines·mm^{-1} grating and 0.76 mm entrance and exit slits allowing a bandwidth of 5 nm. The output intensity at 310 nm was 0.3 mW·cm^{-2}. The optical absorption spectra were measured with a Genesis-5 UV/Visible wavelength spectrophotometer (Milton Roy Inc.) equipped with an IBM-PC. The OA spectra were deconvoluted using a commercial "Peak-Fit" non-linear curve fitting program from Jandel Scientific with Gaussian peak shapes. In addition background subtraction of the OA spectra from irradiated material was also carried out. OA measurements of samples before irradiation were used as background data.

3. Results and Discussion

The preferential population of the 4.3 eV OA band at high levels of dose is shown in **Figure 5** and suggests on statistical grounds that it may be associated with the e-h configuration and glow peak 5a and the 3.84 eV OA band with the e-only configuration and glow peak 5. The motivation for this work, therefore, was to observe the effect of 3.6 eV and 4.3 eV photon bleaches on the TL glow curve in order to provide evidence which either supports or contradicts this correlation. The photon energy of 3.6 eV (instead of

3.84 eV) was chosen as this energy lies outside the range of energies of the 4.3 eV band and could therefore be assumed to excite only the 3.84 eV trapped electrons. At 4.3 eV there is a slight overlap with the 3.84 eV band, however, this could not be avoided. The glow curves following the 3.6 eV and 4.3 eV bleaches are shown in Figure 6. Both glow curves are shifted somewhat to higher temperatures which is probably due to instrumental/measurement factors [10]. The increased intensity of peak 4 following the 4.3 eV bleach (relative to the 3.6 eV bleach) is visually obvious and is supported by the deconvolution analysis shown in Figures 7 and 8. In the case of both bleach energies, a shift of 3-4°C to higher temperatures in the value of the maximum intensity, T_{max}, of the glow peaks was necessary to achieve an optimum fit using the figure of merit (FOM) criterion [11]. The shift to higher temperatures did not, however, significantly influence the relative results of the deconvolution analysis or the interpretation of these results.

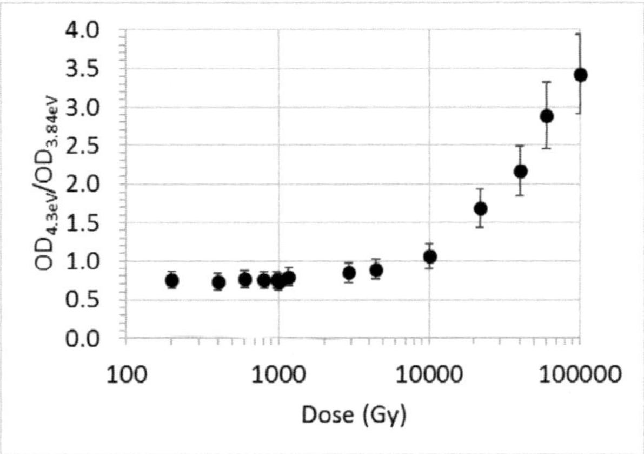

Figure 5. Ratio of the optical density of the 4.3 eV band to the 3.84 eV band as a function of dose. Reproduced from [1].

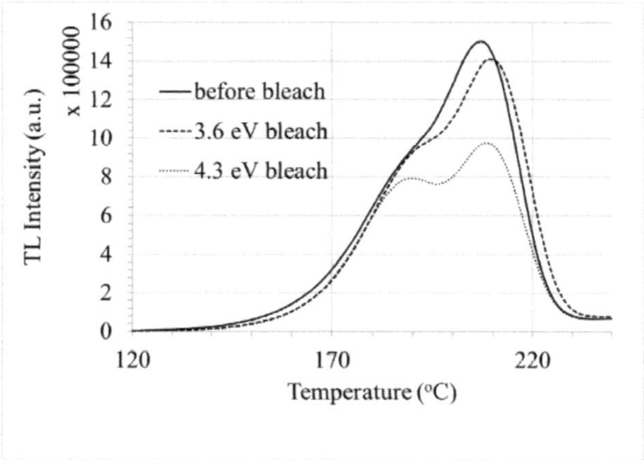

Figure 6. The measured TL glow curves showing the effect of the 3.6 eV and 4.3 eV bleach.

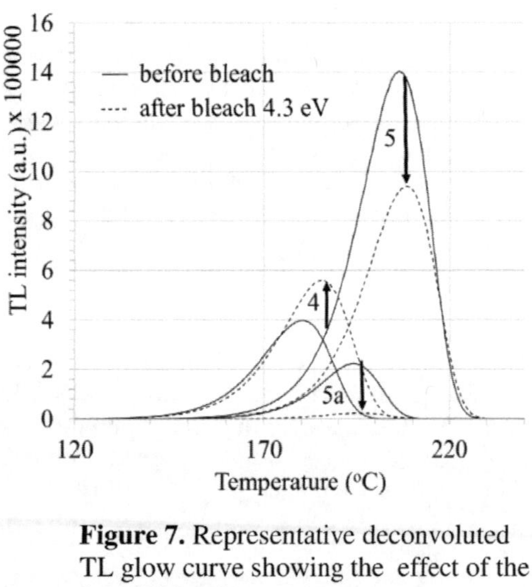

Figure 7. Representative deconvoluted TL glow curve showing the effect of the 4.3 eV bleach.

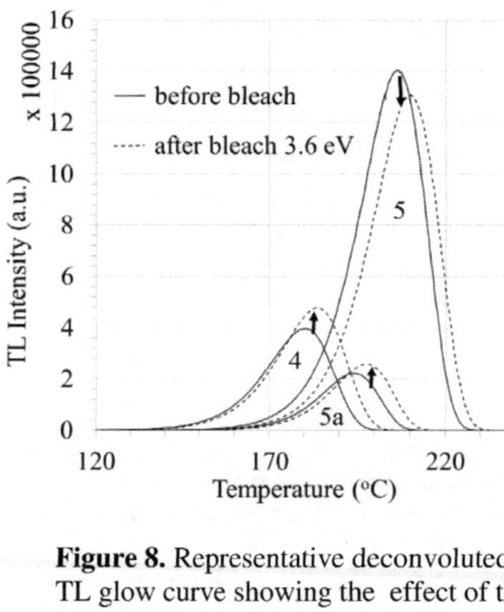

Figure 8. Representative deconvoluted TL glow curve showing the effect of the 3.6 eV bleach.

Figures 7 and 8 show representative (all measurements were carried out on the same experimental conditions) deconvoluted glow curves of peaks 4, 5a and 5 following gamma ray irradiation and the effect of the 4.3 eV and 3.6 eV bleaches on the relative intensities of these glow peaks. Figure 7 shows the effect of the 4.3 eV bleach averaged over four samples. The analysis reveals that this photon energy decreases peak 5 and peak 5a intensity by 32% and 80 % respectively whereas peak 4 increases by 36%. The increase in peak 4 (associated with the h-only configuration) and the decrease in peak 5a (associated with the e-h configuration) can be naturally explained by the removal of the electron from the e-h structure leaving the h-only configuration. The various occupied configurations of the trapping center/luminescent center giving rise to glow peaks 4, 5a and 5 and the effect of bleach are shown in figure 9. Following the 3.6 eV bleach shown in Figure 8, the results are dramatically different. The slight shift in the glow curve maxima was carried out to marginally improve the "goodness of fit" and did not have a significant effect on the results or their interpretation. The intensity of peak 5 is decreased by 19%, however no decrease in peak 5a is observed and peak 4 increases by only 7%.

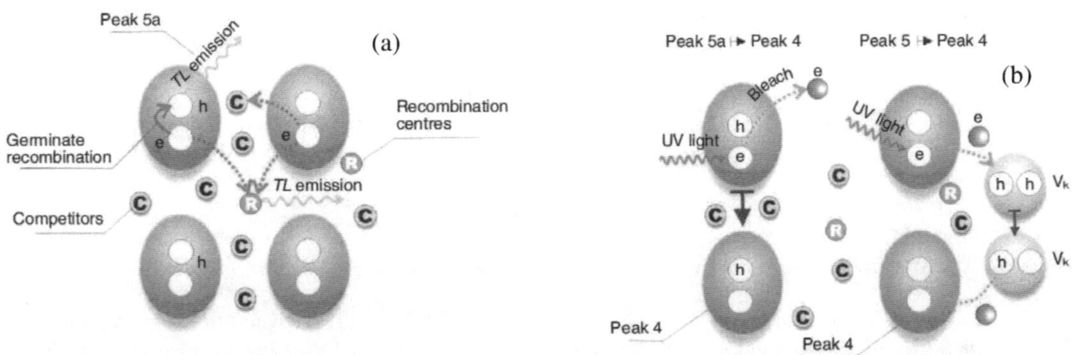

Figure 9. Occupied configurations of the trapping center/luminescent center after irradiation (a) and after irradiation followed by bleach (b).

4. Conclusions

The results strongly support the association of the glow peaks 4,5a and 5 with the h-only, e-h (4.3 eV OA band) and e-only (3.6 eV OA band) configurations respectively. The bleach at 4.3 eV directly transforms the e-h configuration to a h-only configuration leading to the 80% decrease in peak 5a and the 36% increase in peak 4. A similar effect is not observed following the 3.6 eV bleach. Kinetic simulations of the bleach mechanisms will be carried out in the future to investigate the possible effect of V_3-V_K transformation by capture of electrons in the 2-h V_3 center and subsequent hole release [1]. This in order to further investigate the proposed association of the 3.84 eV and 4.3 eV OA bands with these glow peaks.

5. References

[1] Druzhyna S, Eliyahu I, Oster L, Horowitz Y S, Biderman S, Reshes G, and Orion I, 2016 Investigaton of the optical absorption dose response of LiF:Mg,Ti (TLD-100) and the role of V centers in F center (5.08 eV) bleaching *Radiat. Meas. in press*

[2] McKeever S W S, 1984 Optical absorption and luminescence in lithium fluoride-TLD-100 *J. Appl. Phys.* **56** 2883-9

[3] Bos A J J and de Haas T M 1998 Temperature dependent absorption spectrometry on LiF:Mg,Ti *Radiat. Meas.* **29** 349-53

[4] Nail I, Oster L, Horowitz, Y S , Biderman, S and Belaish, Y 2006 Properties of the 4.45 eV optical absorption band in LiF:Mg,Ti *Radiat. Prot. Dosim.* **119** 244-7

[5] Horowitz Y S, Oster L, Satinger D, Biderman S and Einav Y 2002 The composite structure of peak 5a in the glow curve of LiF:Mg,Ti (TLD-100): Confirmation of peak 5a arising from a locally trapped electron-hole configuration *Radiat. Prot. Dosim.* **100** 123- 6

[6] Horowitz Y S, Einav Y, Biderman S, and Oster L 2002 Localised and delocalised optically induced conversion of composite glow peak 5 in LiF:Mg,Ti (TLD-100) to glow peak 4 as a function of post-irradiation annealing temperature *Radiat. Prot. Dosim.* **100** 135-8

[7] Weizman Y, Horowitz Y S and Oster L 2002 Kinetic modelling of the optically stimulated conversion of peaks 5a and 5 to peak 4 in LiF:Mg,Ti (TLD-100) *Radiat. Prot. Dosim.* **100** 131-9

[8] Eliyahu I, Horowitz Y S, Oster L and Mardor I 2014 A kinetic model incorporating both localized and delocalized recombination:Application to the TL dose response on photon energy *J. Lumin.* **145** 600-7

[9] Horowitz Y S, Fuks E, Oster L, Podpalov L, Belaish Y and Ben Shachar B 2007 Advanced multi-stage deconvolution applied to composite peak 5 in LiF:Mg,Ti (TLD-100) *Radiat. Prot. Dosim.* **126** 322-5

[10] Sibony D, Horowitz Y S, and Oster L 2014 The effect of planchet/sample geometry and temperature resolution on precision in LiF:Mg,Ti Dosimetry *Radiat. Meas.* **71** 205-7

[11] Balian H G and Eddy N W 1977 Figure of Merit (FOM), An improved criterion over the normalized chi-squared test for assessing goodness-of-fit of gamma ray spectra peaks *Nucl. Instrum. Meths.* **145** 389-95

IOP Conf. Series: Materials Science and Engineering **169** (2017) 012017 doi:10.1088/1757-899X/169/1/012017

On the real structure of profiled anion-deficient corundum

V I Maksimov[1], **V I Sokolov**[1], **A I Surdo**[2,3], **R M Abashev**[2,3], **E N Yushkova**[1]

[1] Institute of Metal Physics UB RAS, 18 S. Kovalevskaya Str., 620990 Ekaterinburg, Russia

[2] Ural Federal University, 19 Mira Str., 620002 Ekaterinburg, Russia

[3] Institute of Industrial Ecology UB RAS, 20 S. Kovalevskaya Str., 620219 Ekaterinburg, Russia

maximov@imp.uran.ru

Abstract. Profiled Al_2O_3 single crystals grown by Stepanov's method to obtain anion-deficient composition were characterized by neutron diffraction at T=300 K for the first time. Whereas the main structure motif of investigated crystals is checked to be of corundum-type, the scattering pictures of as-grown crystal demonstrate pronounced anomalies being probably indications on substructure forming. However, neutron scanning of synthesised crystal taken after annealing under restoring conditions reveals additional effects associated with displacement type superstructure.

1. Introduction

There is considerable interest in the refractory compounds (mainly oxides) to be adapted for a range of high-temperature and nuclear applications, variations of sintered corundum, or Al_2O_3, belong to large group of these numerous materials [see 1-3 and refs therein]. As an example based on experimentally investigated Al_2O_3 properties, the defect structure of these compounds is of great importance for atomic transport and for response to high-energy radiation [3].

In idealized model, represented by space group R-3c (trigonal syngony) the α-Al_2O_3 structure may be described as a slightly distorted hexagonal close-packed arrangement of oxygen ions (O^{2-}), with interlayer coordinations related to displacements of aluminium atoms. Neglecting the known deviations from the "ideal" close-packing, the aluminium ions (Al^{3+}) are placed in the octahedral sites [4, 5]. Hence the Al^{3+}-ions occupy two-thirds of the interstices in the structure. In actual fact, the angle between in-plane <100> and <010> directions is not truly equal 120^0, the angle difference from ideal packing is about 4.3^0 [see 6 and appropriate refs. therein]. The question about possibilities to form superstructures based on structure vacancies subsystem of real structure of Al_2O_3 is still open.

Profiled anion-deficient single crystals of corundum (α-$Al_2O_{3-\delta}$) grown by Stepanov's method under reducing conditions are widely used for the manufacture of radiation detectors of the type TLD-500 and Al_2O_3:C [7]. There are several models to describe recombination processes underlying radiation-detection base represented by thermo- or optical-stimulated luminescence read-out [8-12], but interpretation of the available experimental information obtained on anion-deficient Al_2O_3 is by no means clear cut to define mechanisms of radiation detection strictly.

The presented work is devoted to neutronographic characterization of structure states of anion-deficient corundum single crystals grown by Stepanov's method; another purpose in view - to check whether structure vacancies of anion-deficient corundum are possible to form any superstructures after restoring anneal of α-$Al_2O_{3-\delta}$ crystals.

2. Experimental

2.1. *The samples*

Two types of corundum single crystals were investigated by scattering of thermal neutrons at T=300 K. The first type (or α-$Al_2O_{3-\delta}$) as grown crystal was with a concentration of anion vacancies ~10^{17} cm^{-3}. The second type crystal obtained by the same synthesis was annealed at T=1970 K for 70 hours in oxygen atmosphere to restore the "natural" structure state of corundum or α-Al_2O_3. In the second case concentration of anion vacancies was less than 10^{12} cm^{-3}. The single crystals were in cylindrical shape. (Their heights and diameters was about 0.4 cm.)

2.2. *Experiment*

The neutron diffraction experiment have been performed on D7b single crystal diffractometer (λ=1.57 Å) at IVV-2M research reactor (Sverdlovsk region, Zarechny, Russia).

3. Results and discussion

Major features of the idealized structure model identified by visible main Bragg peaks of thermal neutron diffraction allow the adequate adjustment of appropriate volume α-Al_2O_3 single crystals for neutron diffraction experiments. As it was checked by neutronographic characterization, corundum structure is the main structure motif of both crystals. Appearance of very weak (003) and (009) reflections, forbidden for this structure, is in character for both investigated crystals and indicates on light deviations of the main structure motif from corundum type by, presumably, faults associated with, firstly, close-packing sequence and, secondly, non-pronounced disarrangements in related cation surrounding.

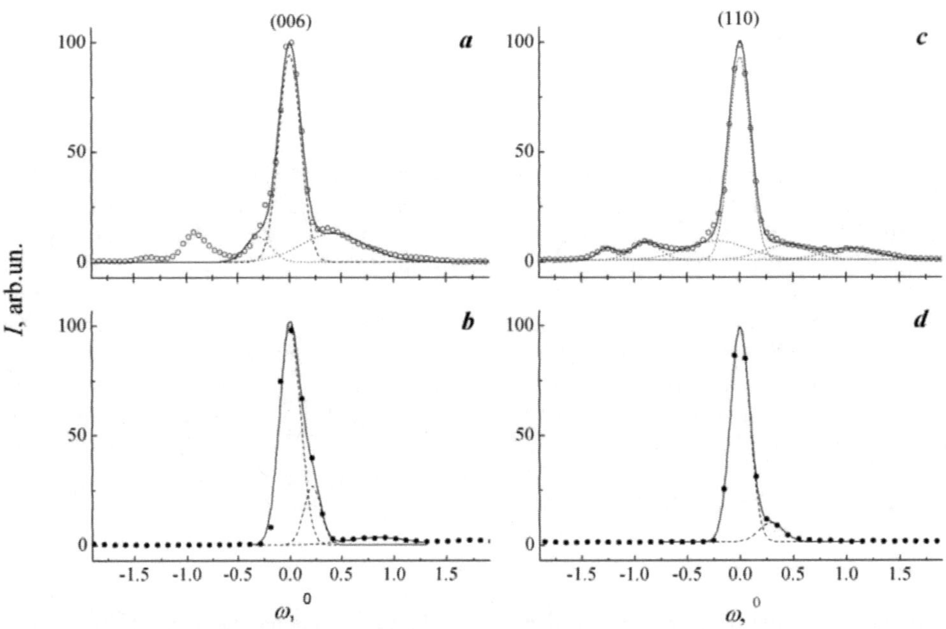

Figure 1. The ω-scans of 006 (a, b) and 110 (c, d) reflections in (100) section: a, c –α-$Al_2O_{3-\delta}$ crystal as grown (○); b, d – the second crystal after annealing (□●). T=300 K.

From the scans of reflection curves (or ω-scans) of the most intensive Bragg reflections, real structure of as grown α-$Al_2O_{3-\delta}$ crystal is characterized by existence of single massive crystallite of corundum structure in presence of large numbers of small fragments, with full volume of them being in commensurability with the volume of the main crystal block. As it is illustrated by fig.1, various scattering pictures in the vicinities of different reflections laying in one section plane point out strong disorientation in subsystem of micro crystallites. In α-$Al_2O_{3-\delta}$ crystal these microfragments, almost of all, are deformed and irregular strains are seemed to be concentrated in their volumes. Data obtained on annealing crystal show suppression of fragmentation indications.

Information obtained by full scanning of (100) section is noteworthy. As it seen from fig2a, turned "sublattice" can be identified in reciprocal lattice of anion-deficient crystal. The "turn angle" value is estimated to be of 4.5^0. Simple analysis put through the "main" and "additional" knots represented by fig.2a shows that intensities of revealed sublattice reflections are few percent in comparison with identical structure reflections. Because of the "turn angle" is close to angle value given by imperfection from distorted atomic arrangement in (001) hexagonal plane, twinning effect seems to be involved in forming of the subsystem of reflections. Therefore additional striking fragmentation may be produced by twinning in α-$Al_2O_{3-\delta}$ crystal. Instead sublattice knots, point system is revealed on scattering picture obtained from annealed sample (fig. 2b).

To compare intensity relations of the "main" lattices of both crystals and sublattice of anion-deficient case, detailed scanning of each structure knot has been performed on both crystals, and sublattice knots presented by fig.2a have been also scanned. It is found, the intensities characterized both main lattices are identical ones, the sublattice intensity relations of anion-deficient crystals do not

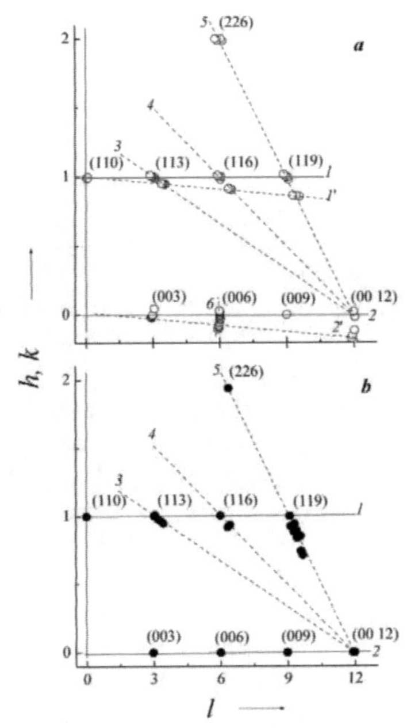

Figure 2. Projections of reciprocal lattice of α-$Al_2O_{3-\delta}$ (a) and α-Al_2O_3 (b) on the (100) plane refined in neutronographic experiment: 1, 2 – knot lines of the "main" lattice; 1', 2' – knot lines of the lattice of "twins"; 3, 4, 5 – lines, along which additional row of knots of annealed sample lattice are detected.

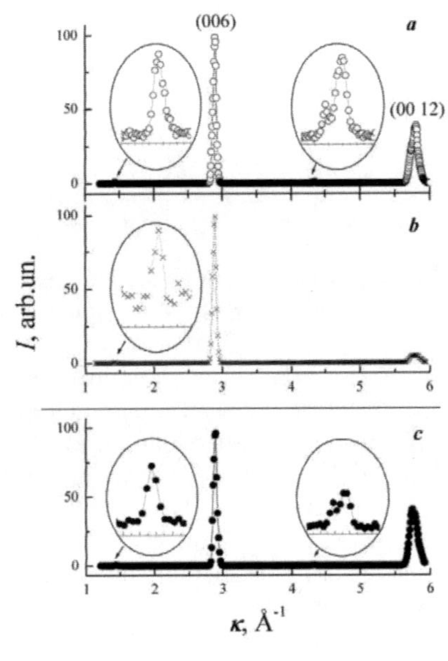

Figure 3. Detailed scans of [001] radial directions: a) α-$Al_2O_{3-\delta}$, main lattice ; b) α-$Al_2O_{3-\delta}$, sublattice; c) data from annealed crystal. Insets represent "forbidden" reflections on enlarged scale.

absolutely repeat analogous characterizations of the main lattice. (The experimental situation is

IOP Conf. Series: Materials Science and Engineering **169** (2017) 012017 doi:10.1088/1757-899X/169/1/012017

illustrated by fig.3, where the relative intensity of (0 0 12) "remote" sublattice reflection is of several times lower in comparison with the some relations for the "main" lattice.) The last fact indicates on existence of substructure characterized by lower symmetry built on the lack of anions.

The knot row placed along 3, 4, 5 lines represented on fig 2b (scattering data from annealed sample) is identified as superstructure knots. Observed very weak reflections is corresponded to wave vectors

$$q = \left(\frac{1}{9}\frac{1}{9}0\right)\frac{2\pi}{a} + \left(0\;0\;\frac{1}{3}\right)\frac{2\pi}{c} \qquad (1)$$

and their second and third harmonics. Bright points of this superstructure are situated along straight lines connecting "main structure" knots, containing places of the "twinning" knots in reciprocal lattice of the anion-deficient crystal. The most demonstrative effect is observed along line 5 on fig. 2b, the detailed scan of which is represented on fig.4.

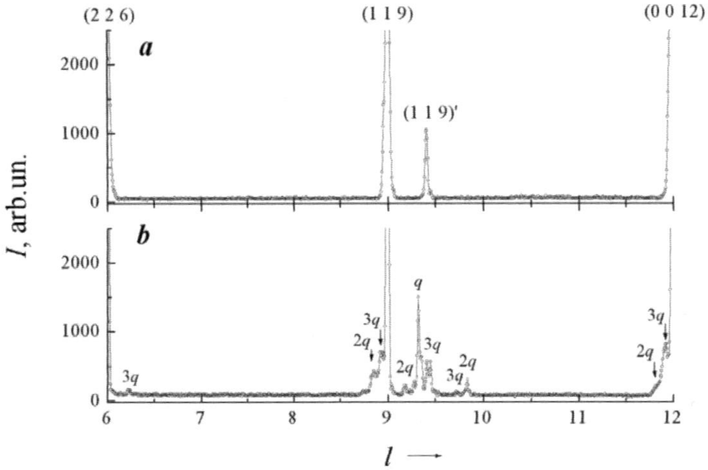

Figure 4. Detailed scan along line connected (0 0 12) and (226) structure knots in (100) section of reciprocal lattice of α-$Al_2O_{3-\delta}$ crystal (a), and α-Al_2O_3 crystal after annealing (b).

Hence these pronounced additional coherent effects are associated with short-wave deformations. As it is pointed out by fig.2b, pronounced components of atomic displacements lay in (11*l*) planes, transverse to <11*l*> directions. Considering appearance of superstructure emerged on atomic displacements, we propose an answer, whether any superstructure can be formed throughout structure vacancies subsystem of α-Al_2O_3 crystal obtained from α-$Al_2O_{3-\delta}$ by annealing. As it is clear from idealized structure model of Al_2O_3, the anion position is surrounded by 4 cations and two structure vacancies gathering triangle-prismatic coordination or irregular tetrahedron built on cations. On this point of view, in simple manner, the data represented by figs 2b and 4 reflect possibilities arisen throughout numerous boundaries of micro crystallites to complete building of close-packed anion sublattice related to deficient case, in which structure restoring tends to form hexagonal axes along four "cubic" space directions simultaneously. Such unstable structure state can be supported by intrinsic defects and localized micro strains. As it is drawn from common formulae (1) giving the precision positions of superstructure knots, real structure displacements tend to form distorted rombohedral structure locally emerged to establish enlarged cell. The "new" hexagonal cell should be built on 18 close-packing anion layers, in which *a* parameter tends to be equal *c*. In this case, the rombohedral angle α seems to be close to 90^0. It reinforces our supposition about disordering of local *c*-axes appearing from turns of Al_4O-complexes, where these turns are accompanied by compensation deformations. Therefore, indications on double and triple superstructure phases seem to involve long-periodicities in real structure of the annealed crystal. Thus, Al- ions should be displaced along local <1-10> hexagonal directions (from filling anion vacancy) to form resulting "pseudocubic" continuum.

4. Conclusion

In this study, the fine features of the structural state of bulk α-$Al_2O_{3-\delta}$ crystals, belonging to perspective detector materials, are considered for the first time by thermal neutron diffraction at T=300 K. In full, neutron scattering picture complicated by fragmented part allows to suppose that anion vacancies, being concentrated mostly in fragments, result in the highest vacancies content on the boundaries of these fragments. The other finding implying twinning effect is presented by additional knots in reciprocal space. Because of intensity relations of quasi-twin indications do not repeat main structure intensity relations, existence of substructure characterized by lower symmetry built on the lack of anions may be assumed. Annealing of α-$Al_2O_{3-\delta}$ crystals in O_2 atmosphere leads to form modulated superstructure of displacement type. To obtain more precision information about atomic displacements in the anion-deficient corundum structure including its restoring sequences, experiments on heightened quality crystals, followed by specific modeling, are needed.

Acknowledgements The work was done at IVV-2M Neutron Material Science Complex within IMP Program "Potok" № 01201463334. The work was partly supported by RFBR (pr. 16-38-00744), UB RAS (pr. 15-2-19) and Act 211 of the Government of the Russian Federation (contract no. 02.A03.21.0006), Center of Excellence "Radiation and Nuclear Technologies".

References

[1] Kotomin E, Tale I, Tale V, Butlers P and Kulis P 1989 *J. Phys.: Condensed Matter* **1** 6777
[2] Kulis P, Rachko Z, Springis M, Tale I and Janson J 1991 *Radiat Eff. Defects Solids* **119-121** 963
[3] Valbis J, Itoh N 1991 *Radiat Eff. Defects Solids* **116** 171
[4] Kelly A and Growes G V 1970 *Crystallography and Crystal Defects* (London: Longman, London)
[5] Kronberg M L 1957 *Acta Metallurgica* **5** 507
[6] Geschwind S and Remeika J P 1961 *Phys. rev.* **122** 757
[7] Zatulovsky L M, Krawiecky D Y, Akselrod M S, Kortov V S, Milman I I, Gottlieb V I, Bichev V R and Schwartz K K 1986 Copyright certificate of the USSR N1340365
[8] Chen R and McKeever S W S 1997 *Theory of thermoluminescence and related phenomena* (Singapore: World Scientific)
[9] Milman I I, Kortov V S and Nikiforov S V 1998 *Radiation Measurements* **29** 401
[10] Surdo A I 2007 *Radiation Measurements* **42** 763
[11] Pagonis V, Chen R and Lawless J L 2007 *Radiation Measurements* **42** 198
[12] Surdo A, Vlasov M, Milman I, Abashev R 2016 *Radiation Measurements* **90** 90

Influence of growth atmosphere on $Ca_3TaGa_3Si_2O_{14}$ single crystals optical properties

N S Kozlova[1], A P Kozlova[1], D A Spassky[1,2] and E V Zabelina[1]

[1] National University of Science and Technology "MISiS", 4, Leninskiy pr.119049, Moscow, Russia
[2] D.V. Skobeltsyn Institute of Nuclear Physics, M.V. Lomonosov Moscow State University, 1(2), Leninskie gory, GSP-1, 119991, Moscow, Russia

zabev@mail.ru

Abstract. $Ca_3TaGa_3Si_2O_{14}$ crystals were grown by the Czochralski technique in different atmospheres. Crystal grown in pure Ar was colourless and in N_2+(1-3)%O_2 – yellowish coloured. In the absorption spectra of yellowish CTGS we observed anisotropy: no significant absorption bands are observed in Z direction, but in X direction we observe absorption bands at 360 nm and at (470-480) nm. In case of colourless CTGS we observed extremely low intensity of the absorption band at ~500 nm, in near UV we observed anisotropy of absorption and absorption band at ~290 nm for X direction. Dichroism degree intensity is extremely low for these crystals, with maximum connected with coloration at 470 nm for yellow crystal and in case of colourless CTGS the main extrema are observed in near UV.

1. Introduction

Structure and properties of the crystals with Ca-gallogermanate structure ($Ca_3Ga_2Ge_4O_{14}$, $La_3Ga_5SiO_{14}$, $La_3Ga_{5,5}Ta_{0,5}O_{14}$ etc.) with point group of symmetry (PGS) 32 are investigated for more than 20^{ty} years since the moment of synthesis of the first representatives of this group in the beginning of 80th of XX century [1, 2], but still new details are being clarified or discovered. For example, the structure of $Ca_3Ga_2Ge_4O_{14}$ was reported in 80th [2] but specified in 2013 [3].

The most investigated and widely used in practice crystals with this structure are $La_3Ga_5SiO_{14}$ (LGS, langasite) and $La_3Ga_{5,5}Ta_{0,5}O_{14}$ (LGT, langatate). It was determined that growth atmosphere significantly influences optical properties of LGS and LGT [4 - 11], so authors of recent manuscript on Ca-gallogermanate family obligatory point out growth conditions. Origin of color centers in these crystals still remains undefined [11]. The defect centers in oxide crystals are studied for many years and trends in their behavior are determined reliably. However, the response of Ca - gallogermanate crystals on certain external treatments differs from other oxide crystals: for instance, annealing of colored LGT or LGS in vacuum leads to bleaching while opposite behavior is generally observed in other oxide crystal [5, 9, 10].

Recently a new effect was observed in colored LGT – dichroism phenomenon [11]. If dichroism is allowed by the structure of the colored crystal then it will manifest itself by the modification of color or color intensity when the crystal is rotated. This effect is most pronounced when observed in the linearly polarized light. In the crystals with PGS 32 the most dichroic direction is along axis 2 of point group of symmetry. Dichroism is studied by the methods of optical spectroscopy and may be observed for absorption bands both in visible and UV or IR regions. This phenomenon may be a slight or pronounced, for example, in LGT difference between two transmission spectra obtained in the same direction with the rotation of the sample at 90° reaches about 11% at the absorption band ~480 nm. [12] Thus, dichroism may be quite significant effect and should be taken into account.

Nowadays more and more compounds with Ca-gallogermanate structure are synthesized. One of them, $Ca_3TaGa_3Si_2O_{14}$ (CTGS, catangasite) is very perspective for both high-temperature piezoelectric applications and non-linear optics [13, 14]. At 500 °C CTGS possess high resistivity with value over 10^9 Ω·cm and its $R{\cdot}C$ (R – electrical resistance, C - capacitance) is 2,36, that is approximately by two orders of magnitude higher than that of LGS [13]. These characteristics together with high thermal stability of dielectric and piezoelectric properties reinforces CTGS potential for high temperature and low frequency applications [13]. Moreover, CTGS is a new laser frequency-doubling crystal [14] that may enrich very limited number of multi-functional self-frequency doubling nonlinear laser crystals and expand areas of their applications.

Compounds with Ca-gallogermanate structure crystallize in the P321 space group with formulae $A_3BC_3D_2O_{14}$, where A is decahedron position, B – octahedral position, symmetry C and D – large and small tetrahedral positions [2]. The crystal structure is characterized by the presence of two layers, which are perpendicular to the axis of 3 order. One layer contains the decahedral and octahedral positions, another – two types of tetrahedral positions.

Ca-gallogermanate compounds are divided into ordered, when all structurally distinct sites are occupied by dissimilar atoms, and disordered, when one type of position is statistically shared by two types of atoms [15]. LGT and LGS are referred to disordered type, and CGTS to ordered one (table 1).

Table 1. Occupancy of the positions by the elements in LGT, LGS and CTGS.

Crystal	Type	Elements' Positions			
		A	B	C	D
$La_3Ga_{5.5}Ta_{0.5}O_{14}$ [16]	disordered	La^{3+}	½ Ga^{3+} ½ Ta^{5+}	Ga^{3+}	Ga^{3+}
$La_3Ga_5SiO_{14}$ [2]	disordered	La^{3+}	Ga^{3+}	Ga^{3+}	½ Si^{4+} ½ Ga^{3+}
$Ca_3TaGa_3Si_2O_{14}$ [17]	ordered	Ca^{2+}	Ta^{5+}	Ga^{3+}	Si^{4+}

Since LGS, LGT and CTGS belong to different structure types the origin of the defect centers in this crystals may also differ.

For successful optical application of CTGS it's important to obtain values of crystals optical characteristics with high accuracy. Also exact studies of CTGS optical properties are important to clarify the origin of defect structure and mechanisms of its formation. This will be helpful both for piezo- and optical applications.

Here we present the results of our studies of the optical characteristics of CTGS crystals, grown in different atmospheres taking into account the dichroism effect.

2. Materials and methods

CTGS crystals were grown by the Czochralski technique in an atmosphere of pure Ar or in N_2 with the addition of oxygen (1-3)%. The studied samples were in the form of cubes or rectangular parallelepiped, with two faces oriented perpendicular to the crystallographic X-axis (the axis of order 2) and Z-axis (the axis of order 3). Crystal grown in Ar was colorless, in N_2+(1-3)%O_2 – yellowish colored. All studied crystals were transparent without cracks or visible inclusions.

The samples were measured in the Accredited testing laboratory "Single Crystals and Stock on their Base" ("SCSB") of the National University of Science and Technology "MISiS" Moscow, Russia, by spectrophotometry method in the wavelength range of 250–800 nm using the spectrophotometer "Cary 5000" ("Agilent Technologies") with the automatic universal measurement accessory (UMA).

3. Influence of the growth conditions on the optical properties

The effect of the growth atmosphere on the color and transmission spectra of LGS, LGT and CTGS crystals was reported [5-7, 11, 13, 14]. First of all it is necessary to describe how growth condition influence on the coloration of these crystals.

LGT is usually grown in pure Ar, Ar with admixture of O_2, or in air [5, 11], CTGS in N_2 with admixture of O_2 or also in air [13, 14, 18] and LGS is grown in Ar or N_2 with admixture of O_2, in air [6, 7, 19, 20] in Pt or Ir crucibles.

According to [5, 11] LGT coloration enhances with increase of oxygen concentration in growth atmosphere (from colorless at Ar):

$$(\mathrm{Ar, Ir}) \rightarrow (\mathrm{Ar} + (0.5\text{-}1)\%\ \mathrm{O_2, Ir}) \rightarrow (\mathrm{air}\ (\sim 20\%\ \mathrm{O_2}), \mathrm{Pt}) \rightarrow (\mathrm{Ar} + 2\%\ \mathrm{O_2, Ir}) \quad (1)$$

As for LGS it's difficult to create equation similar to (1) because the exact oxygen concentration wasn't specified in the papers, never the less authors state [6, 7] that oxygen concentration influences color of LGS as well.

Though in [13] authors approve that CTGS color doesn't depend on growth atmosphere (pure Ar or N_2, N_2+1% O_2) and occurs only due to the material of crucible - yellow in Ir and colorless in Pt. However according to the photos of crystals presented in [13, 14] the color of CTGS gets more intensive as follows (from almost colorless in air, Pt):

$$\mathrm{air, Pt} \rightarrow \mathrm{N_2} + (1\text{-}5)\%\ \mathrm{O_2, Ir} \rightarrow \mathrm{N_2} + 1\ \%\ \mathrm{O_2, Ir} \quad (2)$$

Data on absorption bands in these crystals in connection with crucible material allow to create a table 2 [11, 13, 14].

Table 2. LGT, LGS and CTGS absorption bands in connection with the crucible material.

Effect	Object		Wavelength, nm										Ref.
Induce bands	$(OH)^-$		-[a]									2920	[11]
	Ir^{3+}, Ir^{4+}		200-458				-	-	1798	-	1923	-	[11]
Observed bands	LGS	Ir	280-290	323, 327	350-360	385-400	**476[c]-480**	-	-	1851	-	2941	[11]
	LGS	Ir	280-290	n/r[a]	350-360	n/r	**480**	n/r	n/r	1840	n/r	2920	SCSB
	LGS	Pt	280-300	-	350	-	**450-480**	550, 580	-	-	-	2930	[11]
	LGT	Ir	290-300	n/r	360-370	n/r	**480-490**	n/r	n/r	1840, 1860	-	-	[11]
											n/r	2920	SCSB
	LGT	Pt	290-300	n/r	360-370	n/r	**480-490**	n/r	n/r	1840, 1860	n/r	2920	SCSB
	CTGS	Ir	-	-	340	-	**450**	477	1790	-	-	-	[13, 14, 18]
	CTGS	Pt	-	-	-	-	-	-	-	-	-	-	[13]

[a] – not observed or not reported in literature date as far as we know
[b] – not revealed in laboratory SCSB
[c] – absorption band in this wave-length region that is responsible for colour

One can see (table 2) that in case of LGT and LGS the material of the crucible is not the main reason for the coloration. It's interesting that band at 1798 nm that may be induced by Ir^{3+} or Ir^{4+} wasn't reported in LGT or LGS as far as we know, but it was revealed in CTGS. Connections between coloration of CTGS and crucible material probably make sense to be investigated more explicitly.

Transmission spectra of CTGS were presented in [13, 14, 18]. In [13] authors report no absorption bands for CTGS, grown in Pt crucible, and three absorption bands for CTGS, grown in Ir crucible (340, 450, 1790 nm).

The anisotropy of transmission spectra of CTGS was presented in [18]: authors observed no bands in Z direction and one absorption band at 477 nm in X direction. However, the shape of the transmission curves in this work indicates that the thicknesses of the sample in X and Z directions were different.

In our opinion, it is better to present absorption spectra instead of transmission spectra, since absorption spectra take into account the thickness of the sample. Absorption coefficient is calculated by Buger-Lambert law [21]:

$$I = I_0 \cdot e^{-\alpha d}, \quad (3)$$

where I_0 – intensity of incident light, I - intensity of light transmitted through the sample, d – sample thickness.

According to [11] in polar X direction dichroism should be taken into account. This means that the sample should be installed into the sample holder in the way so that the light will propagate along the crystallographic axis X (position (1)), and then the sample is rotated by 90° around the axis of propagation of light (position (2)). For each position (figure 1), the transmission spectrum was recorded.

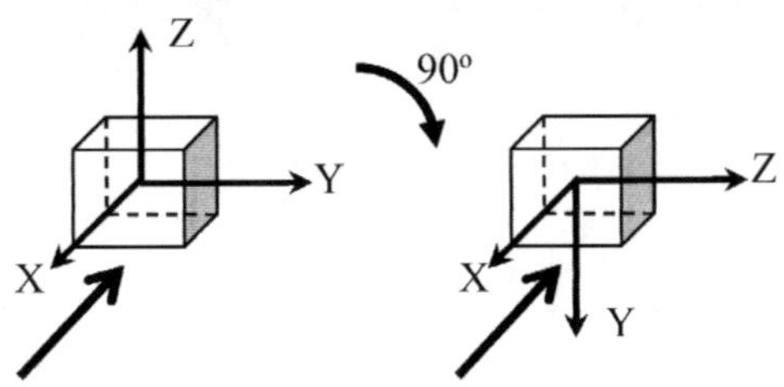

Figure 1. Scheme of dichroism study: left – position (1); right – position (2).

Absorption spectra of CTGS grown in N_2+(1-3)%O_2 are presented in figure 2. Crystal is almost colourless (Figure 2, a) - the intensity of absorption band at visible wave-range is low.

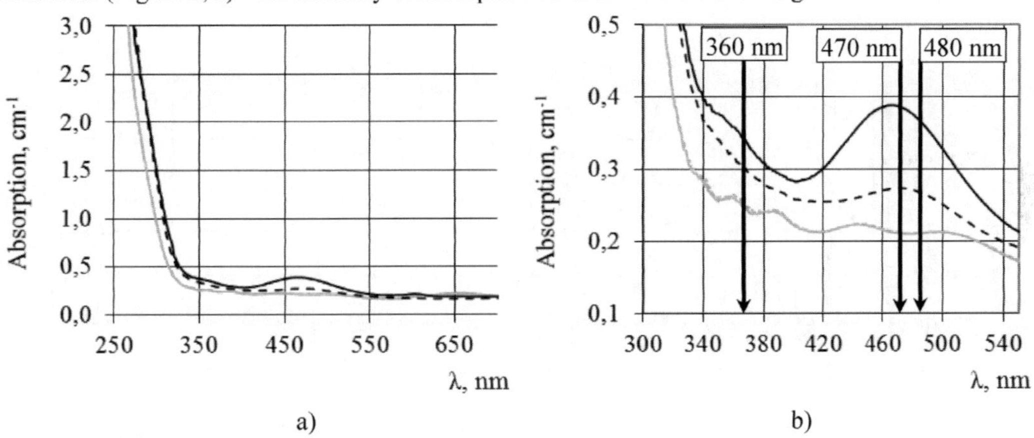

Figure 2. Absorption spectra of CTGS grown in N_2 + (1-3) %O_2 in the wave-range 250-700 (a) and 300-540 (b): position (1) X "— ", Z "—"; position (2) X "- - - -", Z "- - - -".

In the Z axis no absorption bands are observed as well as no dichroism. In the X axis two absorption bands were observed – weak band at ~360 nm and more pronounced at 470-480 nm. These results are in agreement with [18]. The shift of the absorption band maximum with rotation of the sample from ~470 nm at position (1) to ~480 nm at position (2) indicates dichroism.

CTGS grown in Ar is colourless (Figure 3) with extremely low absorption band at ~500 nm. In near UV we observe anisotropy of absorption and absorption band at ~290 nm for X direction.

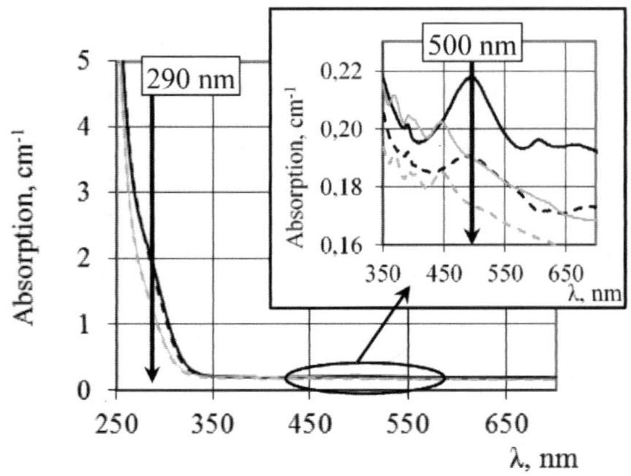

Figure 3. Absorption spectra of CTGS grown in Ar in the wave-range 250-700 (a), 250-350 (b) and 350-700 (c): position (1) X "——", Z "——"; position (2) X "- - - -", Z "- - - -".

We used a usual mathematical formula to gain contrast obtained at the two variations of absorption in positions when light propagates along X direction to calculate the degree of dichroism Δ [22]:

$$\Delta = \frac{D_1 - D_2}{D_1 + D_2} \quad (4)$$

where D_1 is the optical density of the light transmitted through the sample in the position (1); D_2 - in the position (2).

Dichroism spectra of CTGS are presented in figure 4.

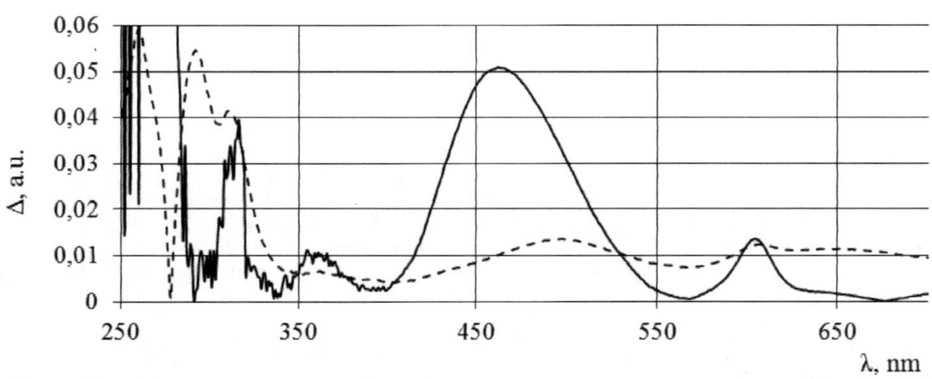

Figure 4. Dichroism spectra of CTGS grown in N_2+(1-3)%O_2 "——" and in Ar "- - - -"

Dichroism degree intensity is extremely low, and it is normal for nearly colorless crystals, though maximum connected with coloration is still at 470 for CTGS grown in N_2+(1-3)%O_2 and in case of CTGS grown Ar the main extrema are observed in near UV.

Dichroism indicates the anisotropy of colour centres.

4. Conclusions

CTGS crystals were grown in N_2+(1-3)%O_2 and Ar atmospheres. Crystal grown in pure Ar was colourless and in N_2+(1-3)%O_2 – yellowish coloured. In the absorption spectra of CTGS grown in N_2+(1-3)%O_2 the anisotropy was observed: no significant absorption bands were observed in Z direction, but in X direction absorption bands at 360 nm and (470-480) nm were observed. In case of

CTGS grown in Ar atmosphere the absorption band at ~500 nm with extremely low intensity was observed, which was accompanied by the anisotropic absorption band at ~290 nm for X direction.

Dichroism degree intensity is extremely low for these crystals, with maximum connected with coloration at 470 nm for N_2+(1-3)%O_2 CTGS crystal and in case of Ar CTGS crystal the main extrema are observed in near UV region.

Acknowledgement
The paper is a part of the research done within the projects RW №278 subject № 3035023 of Ministry of Education and Science of the Russian Federation, № K3-2016-037 of Ministry of Education and Science of the Russian Federation in the framework of Increase Competitiveness Program of NUST "MISiS".

References

[1] Belokoneva E L and Belov N V 1981 *Dokl. Akad. Nauk SSSR* **260(6)** 1363
[2] Kaminskii A A 1986 *Physics and Spectroscopy of Laser Crystals* (Moscow: Nauka)
[3] Dudka A and Mill' B V 2013 *Crystallography Reports* **58(4)** 593-602
[4] Uda S and Buzanov O A 2000 Journal of Crystal Growth **211(1-4)** 318-24
[5] Buzanov O A, Zabelina E V and Kozlova N S *Crystallography Reports* 2007 **52(4)** 691-6
[6] Wang Z M, Yu W T, Yuan D R, Wang X Q, Xue G, Shi X Z, Xu D and Lv M K 2003 *New Crystal Structures* **218(JG)** 421-2
[7] Wang J, Yin X, Zhang S, Kong Y, Zhang Y, Hu X and Jiang M 2003 *Optical Materials* **23** 393-7
[8] Spassky D A, Brik M G, Kozlova N S, Kozlova A P, Zabelina E V, Buzanov O A and Belsky A *Journal of Luminescence* 2016 **177** 152-9
[9] Dubovik M F, Katrunov R A and Korshikova T I *IEEE International Frequency Control Symposium* 1995 **95** 638-41
[10] Tyunina E A, Kuz'micheva G M, Zaharko A and Dubovskii A B *Vestnik MITHT* 2010 **5(5)** 27-35
[11] Kozlova N S, Buzanov O A, Zabelina E V, Kozlova A P and Bykova M B *Crystallography Reports* 2016 **61(2)** 275-84
[12] Kozlova N S, Buzanov O A, Kozlova A P, Zabelina E V, Shayapov V R, Siminel N A and Spassky D A 2016 *Radiation and Applications* in press
[13] Fu X, Villora E G, Matsushita Y, Kitanaka Y, Noguchi Y, Miyayama M, Shimamura K and Ohashi N 2016 *Ceramic Society Japan* **124(5)** 523-7
[14] Chen F, Yu F, Hou S, Liu Y, Zhou Y, Shi X, Wang H, Wang Z, Zhao X 2014 *CrystEngComm* **16** 10286-91
[15] Heimann R B and Hengst M 2004 *Advanced Engineering Materials* **6(7)** 588-94
[16] Takeda H, Sugiyama K, Inaba K, Shimamura R and Fukuda T 1997 *Japanese Journal of Applied Physics* **36(2, 7B)** L919-21
[17] Wang Z M, Yu W T, Yuan D R, Wang X Q, Xue G, Shi X Z, Xu D and Lv M K 2003 *New Crystal Structures* **218(JG)** 421-2
[18] Shi X, Yuan D, Wei A, Wang Z and Wang B 2006 *Materials Reashers Bulletin* **41** 1052-5
[19] Buzanov O A, Naumov A V, Nechaev V V and Knyazev S N 1996 *Proc. IEEE Int. Freq. Control Symp* 131-6
[20] Law B, Appleby J R G , Shimamura K, Takeda H, Kohno T and Fukuda T 1996 *Journal of Crystal Growth* **163** 388-92
[21] Landsberg G S 1976 *Optics* (Moscow: Nauka)
[22] Tudor T and Manea V 2011 *Journal of Modern Optics* **10(58)** 845-52

Spatial distribution of $LiNbO_3$ single crystals optical properties changes after redox high-temperature treatment

D Sugak[1,2], I Syvorotka[2], O Buryy[1], U Yachnevych[1], I Solskii[2], N Martynyuk[1], Yu Suhak[3], G Singh[4], V Janyani[4] and S. Ubizskii[1]
[1] Lviv Polytechnic National University, Lviv, Ukraine
[2] Scientific Research Company "Carat", Lviv, Ukraine
[3] Clausthal University of Technology, Goslar, Germany
[4] Malaviya National Institute of Technology, Jaipur, India

E-mail: crystal@polynet.lviv.ua

The changes in optical properties of bulk $LiNbO_3$ crystals were investigated depending on the diffusion depth. For this purpose the samples were isochronously annealed in hydrogen, subsequently the reduced samples were isochronously annealed in air. The investigations were held in different crystallographic directions. The experimentally determined distribution of depth-dependent additional absorption in reduced crystals was approximated basing on the oxygen vacancies diffusion model and solution of differential equations set describing the quasichemical reaction of bipolarons formation due to the oxygen loss in $LiNbO_3$ crystal.

1. Introduction

High temperature annealing of lithium niobate ($LiNbO_3$, LN) in vacuum, reducing and oxidizing environments, in the presence of metal ions, in proton-rich liquids etc is one of the techniques, which is extensively used to create LN-based elements for acoustoelectronics, optoelectronics and integrated optics [1-3]. However such thermochemical treatments alter the optical and electrical properties of LN. These changes are based on ions diffusion processes out of the crystal into the environment and vice versa, that affect the chemical composition of LN and its defect subsystem state. The diffusion processes in $LiNbO_3$ are studied in a significant amount of works, for example [4-13].

Since in most cases diffusion surface layers with several microns thickness are sufficient for practical applications, the changes in chemical composition of LN layers in depth (and, consequently, changes in their properties) caused by the diffusion processes, are mainly studied by mass spectroscopy of secondary ions (SIMS) (see. e.g. [5]). This technique allows determining the changes only in micron-range and the measurements are performed in the direction of diffusion. Only in [6, 7] the changes in the optical properties of bulk congruent LN samples with about 3 mm thickness, treated by VTE technique, were investigated by means of second harmonic generation conditions analysis and spectrometry in the direction, perpendicular to the direction of diffusion.

Therefore the targeted investigations of coloration changes in LN crystals, caused by redox annealing, depending on the distance from crystals surface (for a few millimeters deep into the crystal) were not performed. Thereby it is not known how the coloration processes occur at different temperatures in crystals volume during redox treatment.

Furthermore, the diffusion mechanisms during thermochemical treatments are also not completely clear. In [8] it is considered that the main diffusion mechanism in congruent $LiNbO_3$, which contains no impurity ions, is the diffusion of lithium and lithium vacancies. In [9] it is assumed that both lithium and niobium diffusion processes take place simultaneously. Authors in [10, 11] consider the oxygen diffusion as the main mechanism and in [4, 12] – the diffusion of oxygen and lithium. Moreover, the known data about anisotropy of the diffusion processes is also controversial. For example in [3] the lithium diffusion is considered as isotropic, however in [13] it is assumed that the lithium diffusion coefficient in Y-direction is greater than in Z-direction. Thus, despite the fact that understanding of the diffusion mechanisms during thermochemical treatments in nominally undoped LN is vital for determining the nature of changes in the crystals, the results of existing works are contradictory. Therefore the aim of this work is the investigation of changes in optical properties with the depth of diffusion in different crystallographic directions of bulk $LiNbO_3$ crystals under the influence of isochronous annealing in H_2 and subsequent isochronous annealing of reduced samples in air. The mathematical simulation for the obtained dependences of properties changes is proposed.

2. Experimental details

All investigated samples were cut from the congruent LN crystals, grown at SRC "Carat" [14]. The samples with dimensions 7(X)×15(Y)×32(Z) mm were sealed in ampoules, filled in with pure H_2 (pressure 0.5 bar at room temperature) and then annealed for 1 our at 400, 500, 600 and 700 °C. The hydrogen was chosen due to the fact that coloration occurs most effectively in this case and additional absorption spectrum has in general a similar structure and nature as after annealing in vacuum.

Five samples, annealed in H_2 at 600 °C were chosen for further treatment in air for 1 hour at 400, 500, 525, 550 and 600 °C respectively. In all cases the heating rate was 5 K / min.

Absorption spectra of the samples after annealing were investigated on UV3600 Shimadzu spectrophotometer (Japan) in the 300-800 nm region with the specially designed device, which allowed scanning by light beam with 150 μm step through the aperture with 100 μm in diameter, and studying the coloring changes with thickness in *X*, *Y*, *Z* directions.

The scheme of crystal cutting after annealing and scan directions of the obtained plates is shown in Fig.1. The cut plates were then polished to the thickness of 1 mm.

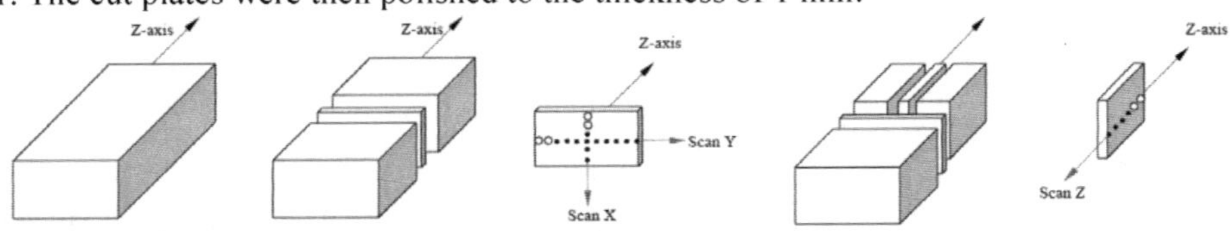

Fig.1. Scheme of the cutting of annealed samples and scan directions of the obtained plates

The changes in structural parameters of reduced LN crystals were determined by the XRD technique on the DRON-3 (Russia) diffractometer in the reflection mode. The comparative studies of the sample surfaces, which were in direct contact with annealing environment and those, which were obtained after cutting from the inner part of the annealed crystal have been performed.

3. Results and discussion

3.1. Annealing in H_2

After annealing in H_2 at 400 °C the uncut samples have dark brown color, after annealing at higher temperatures they become completely opaque. However, this is only an illusion. The coloration is maximal near the surfaces, which were in contact with the annealing environment and gradually decreases as it approaches the center of the crystal. Even after annealing at 700 °C the center of the crystal remains transparent. To characterize the coloration changes depending on the distance from the surface the additional absorption spectra (AA) ΔK were determined at each scan point according to $\Delta K = \frac{1}{d}\ln\frac{T_n}{T_c}$, where d is the thickness of the sample, T_n is the transmission of the sample in the n point, T_c is the transmission in the middle of the sample. In general for *X*, *Y*, *Z* directions and for all studied annealing temperatures it can be assumed that AA spectrum is a superposition of three bands with maxima in the regions of 350-410, 500-520 and 700-750 nm. The relative contribution of these bands to the AA is different and depends on the distance from the surface to the point where the spectrum has been recorded and on the annealing temperature (Fig. 2). The difference in the position of the AA maxima is also observed depending on the crystallographic direction. For example, AA spectra, registered at different points of crystal (scanning along *Z* – axis) after annealing at different temperatures are shown in Fig. 2. As it can be seen from Fig. 2 the shortwave band prevails in the spectrum after annealing at 400 °C. However after annealing at 500 °C the band with the maximum at 500-520 nm is predominant. It is necessary to note that authors in [15] have also observed the similar difference in spectra after reducing annealing of thin LN plates in vacuum. After annealing at 700 °C the absorption in the short-wave region prevails in an area, which is adjacent to the sample surface. Besides the shortwave band and the band with the maximum at 500 nm, the band with the maximum at 700-750 nm is clearly evident in spectrum. Its intensity decreases with the distance from the sample surface and the intensities of the shortwave band and 500 nm band become comparable.

Anisotropy of coloration is illustrated in Fig. 3, which shows the AA change at 510 nm depending on depth along *X, Y, Z* directions after annealing at 700 °C. The deepest coloration is observed in Z-direction to a distance of about 10 mm from the sample surface; in Y-direction the coloration depth is about 6 mm and in X-direction – 4 mm. These differences are apparently related to the crystal structure – along the third order axis of symmetry (Z-direction) there are so-called "channels" that are likely to facilitate the movement of ions in the $LiNbO_3$ lattice. The dependence of the absorption changes with depth is similar also at other wavelengths.

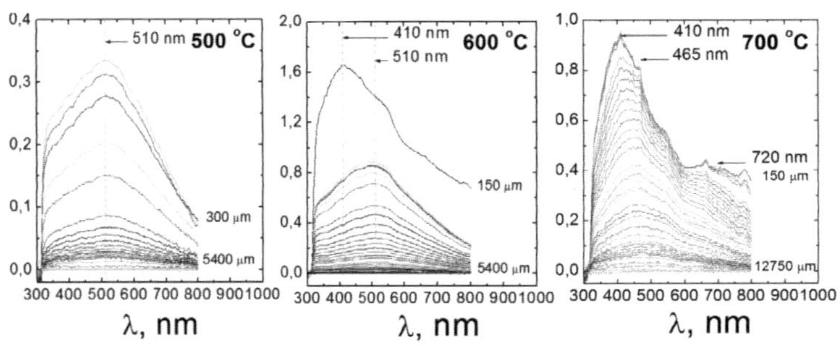

Fig.2. Additional absorption ΔK (see above) spectra of LN after annealing in H_2 at different *T* and different depths from the surface (scanning along Z-direction).

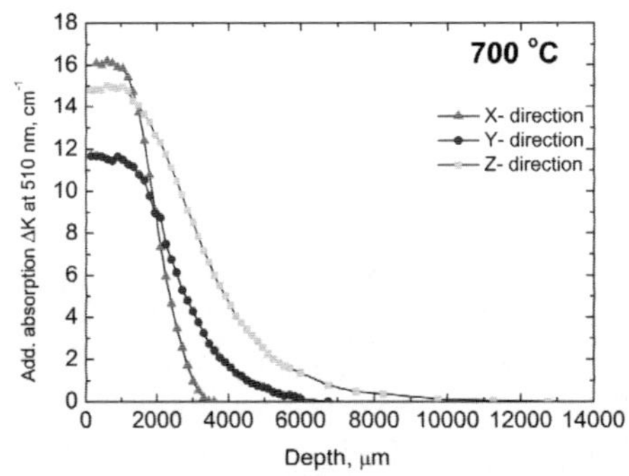

Fig. 3. Change of additional absorption ΔK at $\lambda = 510$ nm in LN depending on depth along *X, Y, Z* directions after annealing in H_2 at 700 °C. Dots - experimental data, solid lines – approximation (see text for details).

X-ray studies of the LN sample show that after annealing in H_2 at 700 °C the crystal structure of the surface, which has been in direct contact with the annealing environment changed (see Fig. 4). Observed reflections correspond to phase $LiNb_3O_8$, which is compared with $LiNbO_3$ depleted Li, according to the ICDD database. The thickness of the $LiNb_3O_8$ layer estimated from the X-ray extinction coefficient is not lower than 25 μm.

3.2. Annealing in air after high temperature H_2-treatment

Oxidizing annealing of LN samples that were previously annealed in H_2 at 600 °C leads to discoloration crystals. The sample plates, cut from the crystals, which were annealed at different temperatures, are shown in Fig. 5. The transparent area with a thickness of less than 100 μm appears near the crystal surface after it was heated to 400 °C. In the temperature range of 500-550 °C the discoloration process occurs more intensively and at 600 °C the sample is almost transparent.

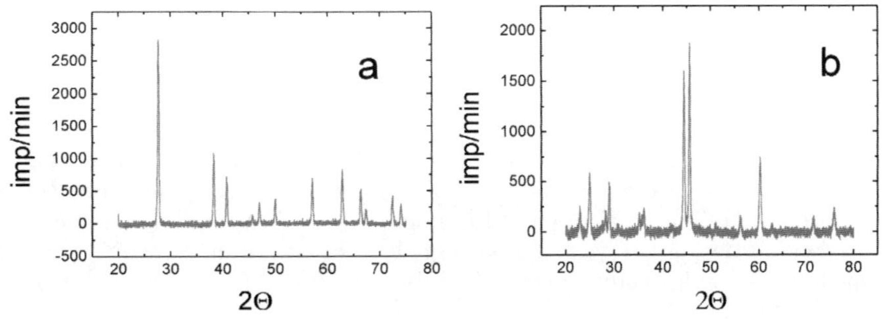

Fig. 4. Diffraction patterns of LN surfaces after annealing in H_2 at 700 °C: *a* – internal surface of the cut sample, $LiNbO_3$ phase; *b* – surface, which was in contact with environment, $LiNb_3O_8$ phase.

Fig. 5. The plates, cut from the crystals, which were previously annealed in H_2 at 600 °C and subsequently annealed in air at 400, 500, 525, 550, 600 °C (the plates are arranged gradually depending on the temperature, the right plate corresponds to 400 °C and the left – to 600 °C)

Thin transparent layer of the sample, treated at 400 °C does not allow measuring the absorption changes with depth by the above described method. Investigations of samples that were annealed at higher temperatures show that absorption in each of the selected spectral range (350-410, 500-520 and 700-750 nm) has the maximum and its position shifts deep into the crystal with the increase of the annealing temperature. As an example, the changes of AA along Z-direction are shown in Fig. 6. The changes were registered at 510 nm after air annealing at 525 °C and 550 °C.

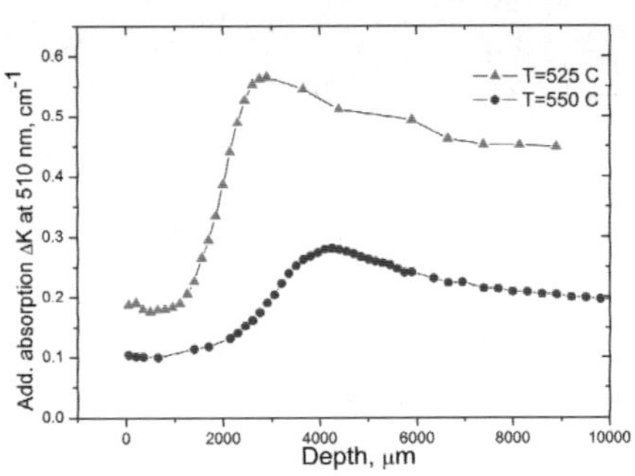

Fig. 6. Changes of additional absorption of LN along Z-axis (λ=510 nm) after air annealing at 525 and 550 °C

3.3. Mathematical description of the $LiNbO_3$ coloration change under redox treatment

As it is seen from the aforementioned results, as well as from the results, obtained previously in the *in situ* mode [16], observed coloration changes caused by change of annealing athmoshere from reducing to oxidizing and vice versa are reversible. Thus, it can be assumed that the oxygen in- and out-diffusion processes underlie the coloration/discoloration mechanisms in $LiNbO_3$ crystals. The model, which were used to describe the depth dependent coloration changes in LN after redox annealing, bases on the following key assumptions: (i) the oxygen diffusion occurs according to the vacancy mechanism; (ii) vacancy (V_O) in-diffusion during reduction leads to the formation of new defects that increases the optical absorption of the crystal; (iii) the defects formation process can be described as the quasi-chemical reaction of the second order.

These assumptions were used to describe the depth distribution of AA at 510 nm, which occurs in LN after reducing at 700 °C (Fig. 3). Most authors, who have studied the optical properties of reduced LN, connect this absorption band with bipolarons formation (bound pair: Nb^{4+} in a regular structural position – Nb^{4+} in the position of Li). Initially the defects, which are formed as the result of oxygen vacancies diffusion to crystal are bound polarons (Nb^{4+} in the structural position of Li) [16, 17], and the formation of bipolarons on their basis is considered to occur instantly for simplification.

The V_O diffusion is described by the second Fick's law:

$$\frac{\partial n_v}{\partial t} = D\frac{\partial^2 n_v}{\partial^2 x}, \tag{1}$$

where n_v is the V_O concentration, D is the diffusion coefficient, x is the distance from the crystal surface in the diffusion direction and t is the annealing time.

On the crystal surface the boundary conditions of the third order are applied:

$$-D\frac{\partial n_v}{\partial \vec{n}} = R(n_v - N_{Vf}), \tag{2}$$

where R is the oxygen mass transfer coefficient from the crystal to the environment, N_{Vf} is the equilibrium V_O concentration after annealing, $\vec{n}$ is the internal normal to the surface.

The reaction of bound polarons formation can be written as following:

$$V + X \leftrightarrow Y, \tag{3}$$

where V denotes an oxygen vacancy, X is the niobium ion in lithium position Nb_{Li}, Y is the bound polaron.

The velocities of the direct and inverse reactions of bound polarons n_y concentration change can be written according to (3) as:

$$\frac{\partial n_y}{\partial t} = -\frac{\partial n_x}{\partial t} = k_f n_v n_x, \qquad -\frac{\partial n_y}{\partial t} = \frac{\partial n_x}{\partial t} = k_b n_y, \tag{4}$$

where n_x is the concentration of Nb_{Li}^{5+} ions, k_f тa k_b are the velocity constants of the direct and inverse reactions in (3). Considering the processes of diffusion and formation/destruction of bound polarons, one gets a set of differential equations:

$$\begin{cases} \frac{\partial n_v}{\partial t} = D\frac{\partial^2 n_v}{\partial^2 x} - k_f n_v n_x + k_b n_y, \\ \frac{\partial n_x}{\partial t} = -k_f n_v n_x \quad k_b n_y, \\ \frac{\partial n_y}{\partial t} = k_f n_v n_x - k_b n_y. \end{cases} \tag{5}$$

To simlify the calculations, all concentrations were normalized on the total concentration of Nb_{Li} ions $N_0 = n_x + n_y$.

Determining of $n_y(x)$ from (4) allows calculation of the absorption change by the formula:

$$A(x) = k_c n_y(x), \tag{6}$$

where k_c is the proportionality factor.

Approximation of experimental dependencies was made basing on (6) by the Levenberg-Marquardt method with the following parameters of approximation: D, R, k_f, k_b, k_c, N_{v0}, N_{vf} (N_{v0} – V_O concentration before annealing). The approximation results for different crystallographic directions are given in Fig. 3 (solid lines) and the obtained values of fitting parameters are indicated in Table 1. As seen from Fig. 3, the results of approximation are satisfactory. Obtained dependences of AA change caused by reducing annealing allowed calculation of V_O diffusion coefficients in *X, Y, Z* directions: $D_x = 3\cdot10^{-6}$, $D_y = 7\cdot10^{-6}$, $D_z = 1{,}2\cdot10^{-5}$ cm^2/s. This result confirms the anisotropy of diffusion processes caused by the peculiarities of the $LiNbO_3$ structure.

Approximation of coloration change curves during oxidation annealing of previously reduced crystals using the same model did not allow obtaining an adequate fit. This may be caused by the fact that the process of oxidation is more complicated due to the lithium loss during the reduction processes and due to the oxygen diffusion from the areas of the crystals, which were not reduced.

Table 1. The obtained values of fitting parameters

Direction	D, 10^{-6} cm^2/s	R, cm/s	$k_f\cdot N_O$, s^{-1}	k_b, s^{-1}	$k_c\cdot N_O$, cm^{-1}	N_{v0}/N_O	N_{vf}/N_O
x	3	0.17	$5.4\cdot10^{-4}$	$2.7\cdot10^{-4}$	17.2	$2\cdot10^{-13}$	13.0
y	7	0.17	$1.9\cdot10^{-4}$	$4\cdot10^{-10}$	11.8	$5\cdot10^{-8}$	13.0
z	12	0.17	$1.3\cdot10^{-4}$	$8\cdot10^{-11}$	15.2	$1\cdot10^{-8}$	13.0

4. Conclusions

Observed reversibility of changes in optical properties after subsequent reducing and oxidizing annealing, allows making an assuption that oxygen diffusion plays a crucial role in coloration/discoloration of $LiNbO_3$.

Comparative X-ray studies showed that the structure of the surface layers corresponds to the $LiNb_3O_8$ phase, while internal surfaces retain the structure of $LiNbO_3$. This can be explained by the partial loss of Li-ions from the crystal surface layers during their reducing thermal treatment.

The AA spectra observed in reduced crystals can be splitted into three bands with maxima in the regions of 350-410, 500-520 and 700-750 nm. The band in 500-520 nm spectral region can be attributed to the absorption of bipolyarons, and the band in 700-750 nm spectral region – to the absorption of bound polarons. The absorption of the short-wave band can probably be attributed to the defects in the anion sublattice that captured electrons due to the oxygen loss in the crystal. The difference in nature of the AA bands is confirmed by the differences in their intensity and spectral position depending on the depth, at which the measurements were performed, crystallographic direction and annealing temperature.

The coloration/discoloration processes in crystals are anisotropic – maximal diffusion depth is observed along the *Z*-direction, minilal – along the *X* one. This can be explained by the peculiarities of the $LiNbO_3$ crystal structure, where movement of ions along the *Z*-direction is facilitated. Experimentally obtained distribution of depth-dependent additional absorption in reduced crystals was successfully approximated basing on the V_O diffusion model and solution of differential equations set describing the quasichemical reaction of bipolarons formation due to the oxygen loss in LN crystal.

The AA maximum on depth-dependent discoloration dependencies after oxidizing annealing of previously reduced LN samples can be explained by two diffusion processes occurring simultaneously in opposite directions – oxygen diffusion from the environment to the crystal and oxygen diffusion from the areas of the crystals, which were not reduced (not colored) after annealing in H_2 at 600 °C.

Acknowledgement. This research was supported by Ministry of Education and Science of Ukraine in the frames of the project EMSh (0116U004134) and the joint Ukrainian-Indian project M/120-2015.

References

[1] Jundt D.H., Fejer M.M. and Byer R.L. 1990 *IEEE J. Quant. Electron.* **26** 135.
[2] Bordui P.F., Jundt D.H., Standifer E.M. [et al.] 1999 *J. Appl. Phys.* **85** 3766.
[3] Nekvindova P., Spirkova J., Cervena J. [et al.] 2002 *Opt. Materials* **19** 245.
[4] Carruthers J.R., Kaminow I.P. and Stulz L.W. 1974 *Appl. Opt.* **13** 2333.
[5] Shi J., Fritze H., Borchardt G. and Becker K.-D. 2011 *Phys. Chem. Chem. Phys.* **13** 6925.

[6] Liang X., Xuewu X., Tow-Chong C. [et al.] 2004 *J. Cryst. Growth* **260** 143.
[7] Wilson R.G., Nowak S.W., Zavada J.M. [et al.] 1989 *J. Appl. Phys.* **66** 6055.
[8] Rahn J., Huger E., Dorrer L. [et al.] 2012 *Phys. Chem. Chem. Phys.* **14** 2427.
[9] Jundt D.H., Fejer M.M., Norwood R.G. and Bordui P.F.1992 *J. Appl. Phys.* **72** 3468.
[10] Birnie III D.P. and Bordui P.F.1994 *J. Appl. Phys.* **76** 3422.
[11] Mehta A. Chang E.K. and Smyth D.M. 1991 *J. Mat. Res.* **6** 851.
[12] Fielitz P., Borchardt G., De Souza R.D. [et al.] 2008 *Solid State Sciences* **10** 746.
[13] Fukuma M. and Noda J. 1980 *Appl. Opt.* **19** 591.
[14] Solskii I.M., Sugak D.Yu. and Vakiv M.M. 2013 *Acta Physica Polonica A* **124** 314.
[15] Akhmadullin I., Golenishchev-Kutuzov V. and Migachev S. 1998 Physics of Solid State **40** 1109.
[16] Sugak D., Zhydachevskii Ya., Sugak Yu. [et al.] 2007 *J. Phys.: Cond. Matter.* **19** 086211.
[17] Schirmer O.F., Imlau M., Merschajann C. and Schoke B. 2009 *J. Phys.: Condens. Matter* **21** 123201.

Amorphisation and recrystallisation study of lithium intercalation into TiO_2 nano-architecture.

M G Matshaba[1], D C Sayle[2], T X T Sayle[1, 2] and P E Ngoepe[1]

[1]Materials Modelling Centre, School of Physical and Mineral Sciences, University of Limpopo, Private Bag X 1106, Sovenga, 0727, South Africa.
[2]School of Physical Sciences, University of Kent, Canterbury, Kent, CT2 7NZ, United Kingdom.

malili.matshaba@ul.ac.za

Abstract. Titanium dioxide is playing an increasingly significant role in easing environmental and energy concerns. Its rich variety of polymorphic crystal structures has facilitated a wide range of applications such as photo-catalysis, photo-splitting of water, photoelectrochromic devices, insulators in metal oxide, semiconductors devices, dye sensitized solar cells (DSSCs) (energy conversions), rechargeable lithium batteries (electrochemical storage). The complex structural aspects in nano TiO_2, are elucidated by microscopic visualization and quantification of the microstructure for electrode materials, since cell performance and various aging mechanisms depend strongly on the appearance and changes in the microstructure. Recent studies on MnO_2 have demonstrated that amorphisation and recrystallisation simulation method can adequately generate various nanostructures, for Li-ion battery compounds. The method was also previously employed to produce nano-TiO_2. In the current study, the approach is used to study lithiated nanoporous structure for TiO_2 which have been extensively studied experimentally, as mentioned above. Molecular graphic images showing microstructural features, including voids and channels have accommodated lithium's during lithiation and delithiation. Preliminary lithiation of TiO_2 will be considered.

1. Introduction

Titanium dioxide is an oxide that is chemically stable, inexpensive, nontoxic and photocatalytic [1]. It has many applications including electrochemical storage for rechargeable lithium ion batteries (LIBs) [2, 3]. Much as the TiO_2 nanoparticles introduce good electrochemical performance of anodes for LIB they also have limitations. It is well known that nanoparticles possess high surface energy compared with the bulk counterpart and incline to aggregate either during synthesis or galvanostatic cycling. The aggregation of nanoparticles decreases the surface area and increases the difficulty for electrolyte solution diffusion within aggregates reaching the surface of particles; it results in the reduction of the total storage energy. Consequently, a better way to avoid this situation is to prepare nanoporous TiO_2 [4]. Nanoporous TiO_2 have been studied by several authors [5, 6, 7, 8] and the micrometre-sized mesoporous materials have characteristic grains as well as pores nearly in the same scale. Electrodes of mesoporous materials for lithium batteries have short transport lengths for Li^+ ions due to their nano-sized grains (10–20 nm), and easy access for electrolytes due to their nanopores (5–10nm). Such mesoporous materials have high packing densities unlike nanopowders, nanowires, nanorods and nanotubes. Despite such advantages, electronic conduction over micrometer-sized particles limits the rate performance of mesoporous materials.

Ren et al. (2010) [9] synthesized an ordered 3D mesoporous anatase using a hard template and investigated lithium intercalation. The structural changes are similar to those observed for nanoparticles, with continuous Li insertion into tetragonal anatase up to $Li_{0.05}TiO_2$, then a two-phase process between anatase and orthorhombic $Li_{0.45}TiO_2$ followed by continuous insertion into the orthorhombic phase up to $Li_{0.96}TiO_2$.
In this work we simulate lithiated nanoporous of TiO_2 to investigate the storage capacity as an anode material inside a LIB. We explore microstructural graphics of our system to see Li ions that move and reside within the nanoporous material. Validation of our model will be accomplished by calculating X-Ray Diffraction patterns (XRDs) which will be compared with the experimental data.

2. Methodology

Potential model and simulation code:
The Calculations are based on the Born model of ionic solids, where Ti^{4+}, Ti^{3+}, Li^+ and O^{2-} ions interact via long-range Coulombic and short range parameterised interactions. A charge is assigned to each atom and the long range attractive interactions between charged ions is balanced by short range repulsive interactions. Model parameters that are used to characterize Li-TiO_2 are specified in table 1 and are taken from Sayle [10] and Sayle [11].
All the molecular dynamics simulations were performed using DL_POLY code [12]. The user manual provides comprehensive analytical descriptions and discussion of the molecular dynamics simulations, force fields, boundary conditions, algorithms and parallelisation methods used in these simulations.

Table 1. Potential parameters describing the short-range potential terms between the component ion species of $LiTiO_2$ used for lithiated TiO_2

interaction (ij)	A_{ij} (Å)	ρ_{ij} (Å)	C_{ij} (eV.Å6)	charge
Ti^{3+}-O^{2-}	18645.84	0.195	22.0	
Ti^{4+}-O^{2-}	16957.53	0.194	12.59	
Li^+-O^{2-}	426.48	0.300	0.00	
Ti^{3+}-Ti^{4+}	28707.21	0.156	16.0	
Ti^{3+}-Ti^{3+}	33883.92	0.156	16.0	1.65
Li^+-Li^+	270000.00	0.143	0.0	0.55
Ti^{4+}-Ti^{4+}	31120.20	0.156	5.25	2.196
O^{2-}-O^{2-}	11782.76	0.234	30.22	-1.098

Generating atomic model
The atomic model for nanoporous architecture was generated using molten TiO_2 nanoparticle precursor. In particular, a particle of TiO_2 comprising 15 972 atoms (5324 titanium and 10648 oxygen atoms), was

cut from the parent bulk material. The nanoparticle was then melted by heating to 6000 K, which is above the melting point for TiO_2, under MD simulation. The molten nanoparticle was then used as the 'building-block' to generate the nanoporous architectures. To capture the (cubic) symmetry of the nanoporous architecture, a simulation cell with cubic symmetry was constructed and a nanoparticle of TiO_2 was positioned at lattice points associated with the primitive cubic Bravais lattice. The size of the simulation cell was then reduced in all three dimensions to enable the nanoparticle to agglomerate in all three spacial directions. NVT simulation was then performed at 2000 K for 3.5 ns to facilitate crystallisation. Periodic boundary conditions were used to replicate the structure infinitely in three dimensions. The low-temperature structural model was then generated by cooling the system to 0 K under NVT MD simulation until the configurational energy converged. The same procedure was performed to create nano-architectures of TiO_2 [13].

Then Li-TiO_2 was generated by randomly inserting 100 Li ions in the amorphous nanoporous TiO_2 at the specific coordinates. Charge compensation in the nanoporous is achieved by changing the titanium (4+ oxidation state) closest to the lithium cation into the Jahn-Teller active titanium (3+ oxidation state). This process is repeated for all the lithium ions inserted in the nanoporous. Amorphous nanoporous structure of Li-TiO_2 is depicted in figure 1 below. Oxygen atoms are red coloured, Titanium are grey and lithium ions are represented by yellow balls.

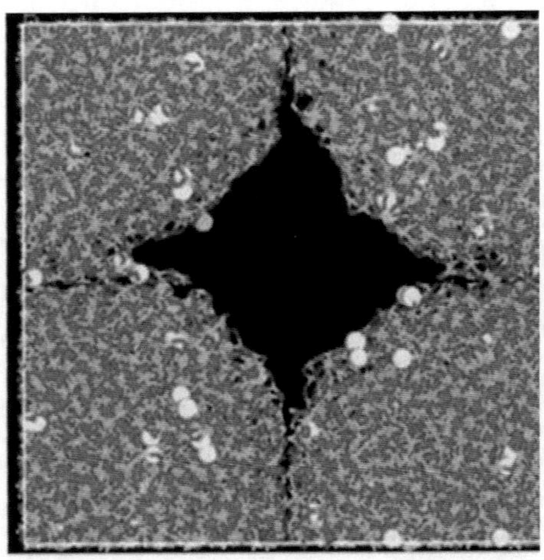

Figure 1. Snapshot of lithiated amorphous nanoporous structure of TiO_2 with 100 Li ions.

3. Results

We performed a molecular dynamics simulation using the NVT ensemble, at a temperature of 2000 K for 700000 steps with a time step of 0.005 ps, in order to recrystallise a lithiated nanoporous of TiO_2. A plot of configuration energy as a function of time is shown in figure 3, indicating that the nanoporous is fully recrystallised since the configuration energy is constant at the energy of -2.04x 10^5 eV. The sudden decrease in energy from 0 to 0.5 ns indicates the transition from amorphous to a crystalline phase.

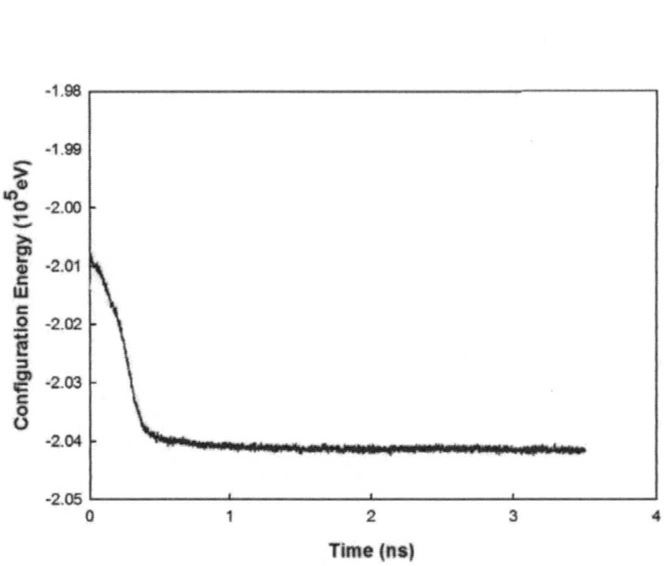

Figure 2. Configuration energy as a function of time for nanoporous TiO_2.

Superimposed radial distribution functions (RDFs) of the amorphous and recrystallised nanoporous structure of Li-TiO_2 are illustrated in figure 3 below. They clearly indicate that our system is recrystallised since the RDFs of recrystallised material shows some sharp picks than the amorphous structure.

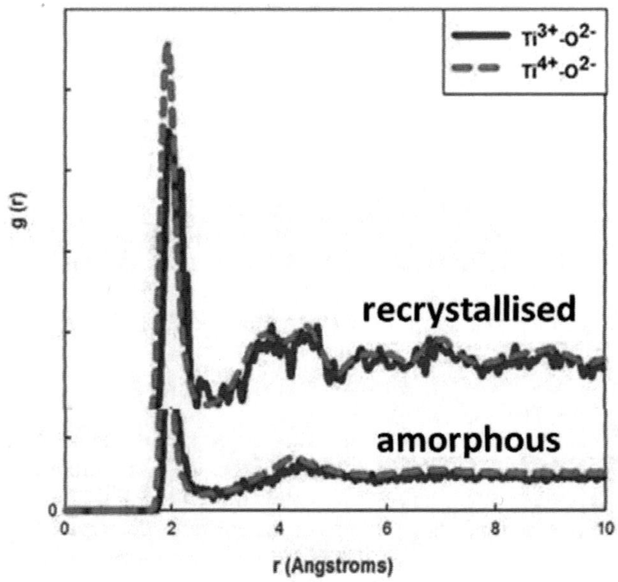

Figure 3. RDFs of lithiated nanoporous of TiO_2 before and after recrystallisation.

The recrystallised lithiated nanoporous of TiO_2 which is depicted in figure 4 was cooled by running molecular dynamics (MD) simulation for 500 ps at 1500 K, followed by 250 ps at 1000 K, 250 ps at 500 K, and lastly 500 ps at 0 K. A similar process was employed in generating the MnO_2 nanosphere [14]. Cooled nanoporous structure of Li-TiO_2 is illustrated in figure 4 and we observe more vacancies next to the channel. Lithium atoms are located closer to the inner edge of the channel in the cooled nanoporous structure of Li-TiO_2 than in the uncooled one; crystalline patterns are more distinct as compared with the ones at higher temperatures.

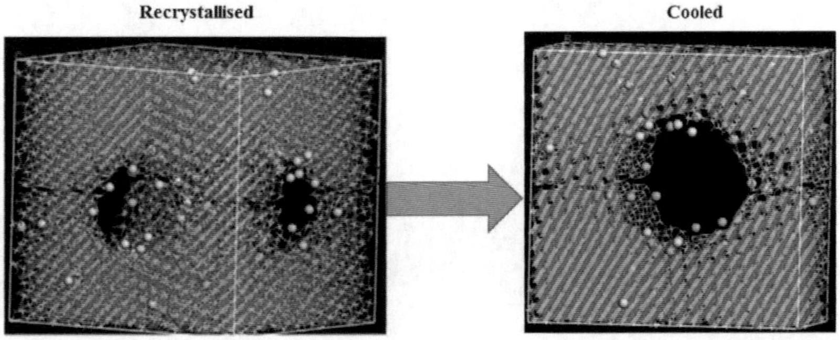

Figure 4. Recrystallised and cooled lithiated nanoporous of TiO_2.

Created microstructure of lithiated nanoporous structure of TiO_2 is shown in figure 5 below. Yellow balls are Li atoms, blue octahedral is the upper layer of Ti^{4+} and the white adjacent correspond to the lower layer of Ti^{4+}. We observe more microstructural characteristics in the nanoporous: rutile, brookite, vacancies, Li's in tunnels, stacking defect, micro-twinning and Li atoms in vacancies. Rutile phase is depicted by the straight tunnels and the brookite phase is characterised by zigzag tunnels. For a nanoporous to be a good candidate of anode material it has to store more Li atoms and be able to have pathways so that it can allow Li atoms to pass through. With Li atoms occupying the lower layer and vacancies it indicate that the nanoporous of TiO_2 can store Li atoms. The material also shows that it can be able to transport Li atoms since it has a clear tunnels and some Li atoms can pass through vacancies.

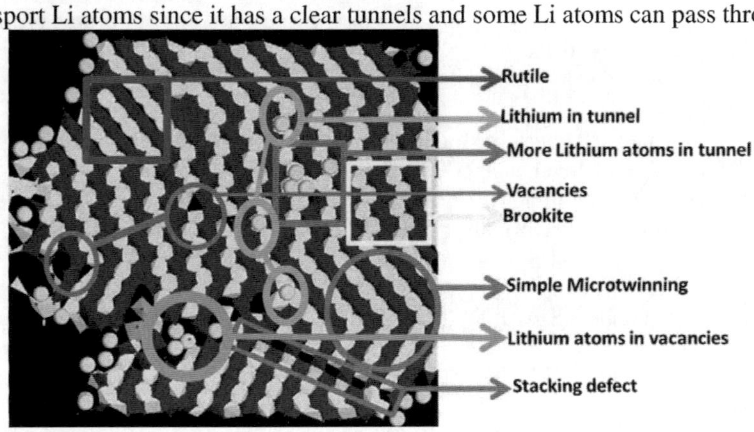

Figure 5. Microstructure of cooled lithiated nanoporous of TiO_2.

On superimposing XRD patterns of the experimental brookite [15], TiO_2: α-PbO_2, rutile polymorphs and the simulated TiO_2 nanosphere, corresponding peaks can be deduced. XRD's are depicted in figure 6 below. At 27, 33 and 37 ° there are sharp peaks that correspond well with the brookite and TiO_2: α-PbO_2 structure, whilst at 42 ° only the simulated, brookite and TiO_2: α-PbO_2 peaks are aligned. The simulated XRD pattern at 50 ° has a smooth curve which tally well with the rutile structure. The peaks at 37 and 57 ° correspond with all polymorphs and the one at 67 ° accords more reasonably with the rutile and TiO_2: α-PbO_2 structures.

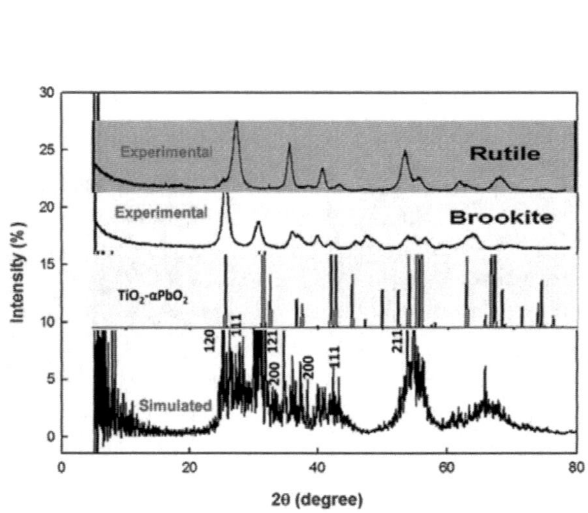

Figure 6. A comparison of simulated TiO_2 nanoporous, calculated TiO_2:α-PbO_2 and experimental[15] XRDs.

4. Conclusion

Molecular dynamics simulation has been used successfully to generate atomistic model of nanoporous Li-TiO_2. The nanoporous of Li-TiO_2 was fully recrystallised and cooled to a 0 K. The recrystallised material was confirmed by the RDFs, which depict sharp picks. Microstructure reveal that the nanoporous of TiO_2 can act as a good anode material since it can store and transport Li atoms during charging and discharging of a battery. XRDs depict the presence of rutile and brookite phases as compared with the experimental data.

Acknowledgements

Calculations were performed at Materials Modelling Centre (MMC), University of Limpopo. This work is supported by Department of Science and Technology, National Research Foundation, Pretoria, the Department of Science and Technology HySA Lithium Ion Batteries and Supercapacitors Project, Pretoria and the Centre for High Performance Computing (CHPC) in South Africa.

References

[1] Liang Y, Gan S, Chambers S A and Altman E I, 2001 *Phys. Rev. B.* **63** 235402-1.
[2] Liu B and Aydil E S, 2009 *J. Am. Chem. Soc.* **131** (11) 3985.
[3] Fujishima A and Honda K, 1972 *Nature* **238** 37.
[4] Hwang K-J, Cho D W, Lee J-W and Im C, 2012 *New J. Chem.* **36** 2094.
[5] Zhou H S, Li D L, Hibino M and Honma I, 2005 *Angew. Chem. Int. Edit.* **44** 797.
[6] Moriguchi I, Hidaka R, Yamada H, Kudo T, Murakami H and Nakashima N, 2006 *Adv. Mater.* **18** 69.
[7] Guo Y G, Hu Y S, Sigle W and Maier J, 2007 *Adv. Mater.* **19** 2087.
[8] Wang Z Y, Liu S Z, Chen G and Xia D G, 2007 *Electrochem. Solid Sate Lett.* **10** (3) A77.
[9] Ren Y, Hardwick L J and Bruce P G, 2010 *Angew. Chem. Int. Ed.* **49** 2570.
[10] Sayle D C and Sayle T X T, 2007 *J. Comput. Theor. Nanosci.* **4** 299.
[11] Sayle T X T, Ngoepe P E and Sayle D C, 2009 *ACS Nano* **3** (10) 3308.
[12] Smith W and Forster T R, 1996. http://www.dl.ac.uk/TCSC/Software/DLPOLY.
[13] Matshaba M G, Sayle D C, Sayle T X T and Ngoepe P E, 2016 J. Phy. Chem. C 120 (26) 14001.
[14] Sayle T X T, Catlow C R A, Maphanga R R, Ngoepe P E and Sayle D C, 2005 *J. Am. Chem. Soc.* **127** 12828.
[15] Dambournet D Belharouak I and Amine K, 2009 *Electrochem. Soc.* **215**, No.1.

Emission Dependent on composition of Si-rich-SiN_X Films obtained by PECVD

J A Jaramillo Gomez[1], T V Torchynska[2], J L Casas Espinola[2], J A Bentosa Gutiérrez[1], L Khomenkova[3] and A Slaoui[4]

[1]Instituto Politécnico Nacional, UPIITA, Ciudad de México, 07340, México
[2]Instituto Politécnico Nacional, ESFM, Ciudad de México, 07738, México
[3] V. Lashkaryov Institute of Semiconductor Physics at NASU, Ukraine
[4]ICube, 23 rue du Loess, BP 20 CR, 67037 Strasbourg Cedex 2, France

Abstract

Silicon-rich silicon nitride films with different stoichiometry were grown on silicon substrate using the plasma-enhanced chemical vapor deposition. The excess silicon content in the films was monitored via a variation of the NH_3/SiH_4 gas flow ratio from 0.45 up to 1.0. Morphology and luminescence properties of the films were studied by means of atomic force microscopy (AFM) and photoluminescence (PL) methods. High-temperature annealing was employed to produce the silicon nanocrystals in the films and to enhance the photoluminescence in the range of 1.6-3.0 eV. The PL spectrum was found to be complex due to the contribution of several radiative channels in emission process. It was determined that their competition leads to the non-monotonous variation of total PL peak position with the increase of the Si excess content. It was observed that the shape of PL spectra depends on an excitation wavelength. The ways to control the PL emission is proposed based on the discussion of the PL mechanism.

Keywords Silicon nanocrystals, silicon nitride, plasma-enhanced chemical vapour deposition, photoluminescence.

E-mail: jantonioj@yahoo.com

1. Introduction

Silicon nano photonics is attracting the significant attention in the field of development of silicon-based photonic devices that allows efficient data transfer at low cost as compared to conventional optoelectronics devices. Bandgap engineering of Si-based materials through the control of the distribution of Si nanocrystals (Si-NCs) offers future applications of these materials in optoelectronics and photovoltaics.

Si_3N_4-based materials have been extensively investigated because of their inherent advantages in the chemical, mechanical and optical properties [1]. The creation of the Si- NCs in a silicon nitride host offers the several key advantages over silicon oxide [2]. Silicon nitride is a more promising matrix for Si- NCs due to its structural stability upon microelectronic fabrication processing. Favorable electrical properties resulting from the lower tunneling barriers allow the better transport of electrons and holes into Si- NCs formed in silicon nitride [3]. In addition, Si- NCs coordinated with oxygen atoms are subject to charge trapping at the interface, which effectively limits the emission quanta from such Si- NCs to less than 2 eV

[4]. Since Si- NCs coordinated with nitrogen atoms do not exhibit the same limitation, the emission has been demonstrated to occur at energies across the entire visible spectrum [3,5,6].

The process of a Si- NCs formation in silicon nitride is also more favorable due to the much lower annealing temperature required for an achievement of bright luminescence compared to silicon oxide [7, 8]. However, the Si- NCs formation in the films has been found to occur in a more complex process, with the appearance of both amorphous and crystalline clusters and a strong dependence on both deposition and processing conditions [3,9,10].

In this work, the morphology and luminescent properties of Si-rich-silicon nitride films grown by plasma-enhanced chemical vapor deposition (PECVD) were investigated versus excess silicon content. The detailed analysis of the results allows getting insight in the nature of emission and excitation processes in the films.

2. Experimental details

Si-rich silicon nitride films were grown by the PECVD technique on the p-type silicon (100) substrates. To achieve different stoichiometry in the films, the NH3/SiH4 (99.9999%) gas flow ratio (R) was varied in the range of R=0.45–1.0 keeping SiH4 flow constant at 14 sccm. The working pressure, the plasma power and the growth temperature were 0.5 Torr, 20 W, and 350 °C, respectively. Thermal annealing at 1100 °C for 30 min in flowing nitrogen was applied to produce the Si-NCs in the films [11].

Morphology and luminescence properties of the films were studied by means of atomic force microscopy (AFM) and photoluminescence (PL) methods. A Nano-Scope IIIa was applied for AFM measurements at room temperature in tapping and covered modes. The PL spectra were excited by either a 325-nm (3.82eV) line of a He-Cd laser or by a 488-nm (2.54eV) line of an Ar+ laser. The power of laser beam was about 80 mW. More details about the PL experiment can be found elsewhere [12, 13]. To investigate the evolution of PL spectra with temperature of measurement in the range of 20-300 K, the samples were mounted in a closed-cycle He cryostat.

3. Experimental results and discussion

AFM images of studied films with different R parameters have been presented in figure 1. The film morphology varies versus R: the film roughness decreases from 13.8-14.8 nm (at R=0.71-1.0) down to 7.8 nm (at R=0.50). It was shown early [14] using the X-ray diffraction study that the SiN_x matrix is characterized by the hexagonal Si_3N_4 phase in annealed films obtained at R=0.7-1.0 with the NC size of 26-30 nm. The NC size in SiN_x matrix falls down to 2.8-3.0 nm (at R=0.63) and, finally, the films became amorphous at R=0.45-0.50 [14]. Thus, the change of roughness detected at AFM study is related, apparently, to the change of crystallinity of the SiNx matrix versus R parameters.

Room-temperature PL spectra of annealed samples demonstrate the variation of their shape with excitation light energy (Fig.2 and Fig. 3). The comparison of Fig.2 with Fig.3 shows that at the higher excitation energy, the broader PL spectrum was observed demonstrating the appearance of asymmetry in the higher energy side (Fig.2). The total PL peak position has shifted also towards the higher energies.

When the films are excited by the 2.54-eV light, only broad PL band was detected and with *R* increasing its peak position shows the tendency to shift gradually from 1.6 eV (*R*=0.45) to 2.2 eV (*R*=0.83) (Fig.3). The PL intensity varied non-monotonously with *R*, demonstrating the highest intensity for the films grown with *R*=0.59-0.67. This behavior of PL spectra indicates the presence of multiple luminescence centers in the films. The spectra can be considered as a superposition of three components peaked at 1.6-1.7 eV (B1) 1.9-2.0 (B2) and at 2.2-2.3 eV (B3) (Fig.4a).

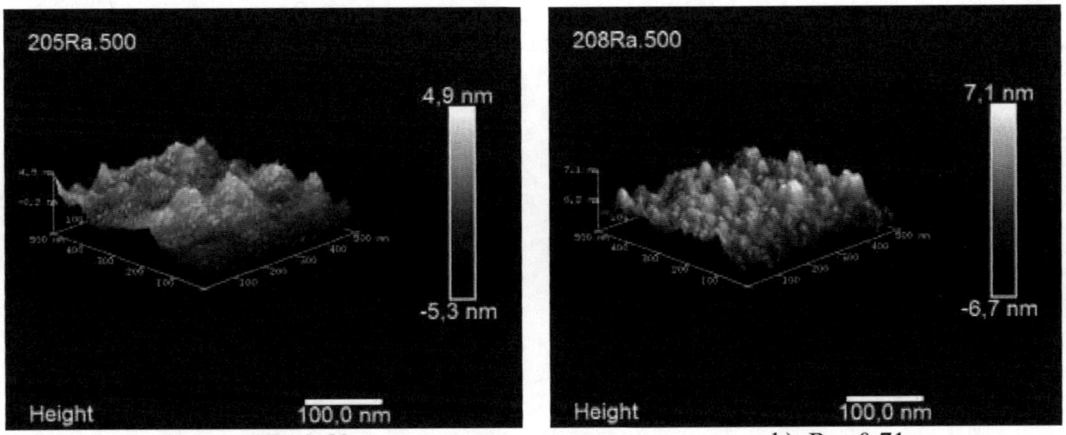

a)R=0.50 b) R = 0.71

Figure 1 AFM images obtained in the films prepared at different R parameters.

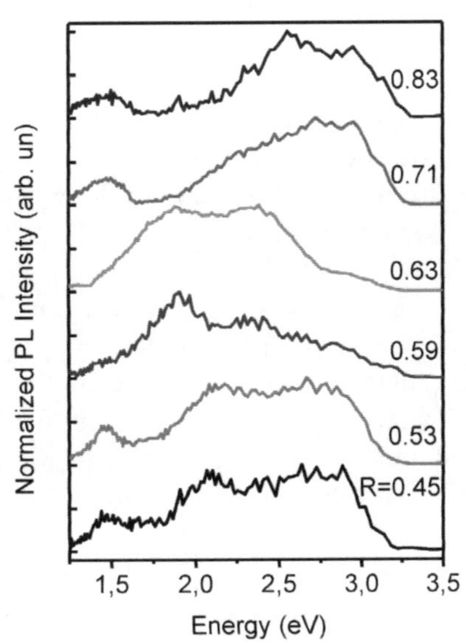

Figure 2 PL spectra of annealed films versus R measured at 3.82-eV excitation light energy. The spectra were normalized on the intensity of their maximum.

The detection of PL spectra under 3.82-eV excitation light supports this assumption. The PL spectra became more complex at higher excitation energy demonstrating also the highest intensity for the films grown with R=0.59-0.67. The total PL peak position does not demonstrate a systematic shift with R varying (Fig.2). The analysis of PL spectra has revealed that all spectra excited by 3.82-eV light can be decomposed on four PL bands with the maxima situated at 1.7-1.8 eV (B1), 2.0-2.2 eV(B2), 2.4-2.5 eV (B3) and 2.9-3.0 eV (B4) (Fig.2 and Fig.4b). It is obvious that three former components are the same as those observed

IOP Conf. Series: Materials Science and Engineering **169** (2017) 012021 doi:10.1088/1757-899X/169/1/012021

under 2.54-eV excitation light described above that demonstrate as well the slight high energy shift of the PL peak position with excitation energy increasing.

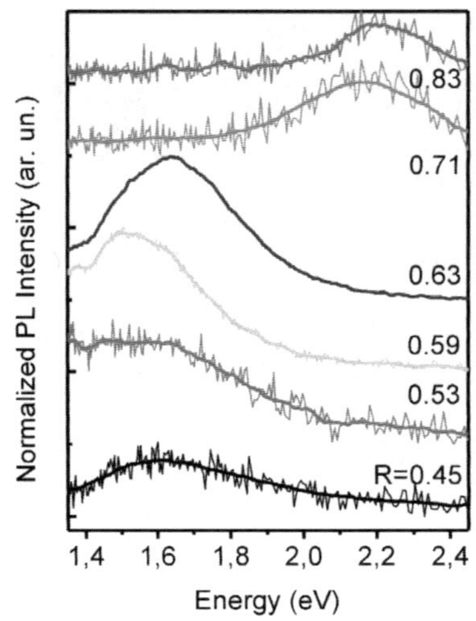

Figure 3 PL spectra of annealed films versus R measured at 2.54-eV excitation light energy. The spectra were normalized on the intensity of their maximum.

The investigation of PL temperature dependences (Fig.5) has showed that only the PL band B1 demonstrates the shift of the PL peak position with cooling similar to the shrinkage of the Si band gap (Fig.5). The B1 peak position at 20K was found to be at about 1.96 eV and its shift to 1.91eV at 300K. Simultaneously, the other PL peaks B2, B2 and B4 do not change their peak position with cooling. Taking into account the spectral position of the PL component B1and its behavior with temperature variation, one can assume that this PL emission is due to the exciton recombination in the Si-NCs [15].

However, one cannot rule out that with cooling the intensity of the B2 component increases synchronously with the intensity of B1 component. This fact can be explained by the close location of corresponding emitting centers to the Si-NCs interface that allows an efficient energy transfer between them to be realized. At the same time, based on the model proposed in [16, 17], the B3 (2.4-2.6eV) and B4 (2.9-3.0 eV) PL components can be ascribed to the Si dangling bonds (so called K^0-centers, ~2.40 eV) and the nitrogen dangling bond levels (N_4^+ and N_2^0, ~3.0 eV). These latter can be excited only by the high-energy excitation light (3.82 eV) as demonstrated in Fig.2. Thus the variation of excitation energy and the temperature of measurements allowed us to detect the different radiative centers in the films. It is seen that for the films fabricated with $R \leq 0.56$ and $R \geq 0.63$ the main contribution into PL spectra is given by the host defects, whereas for R=0.56-0.63 the Si-NCs form the main radiative channel.

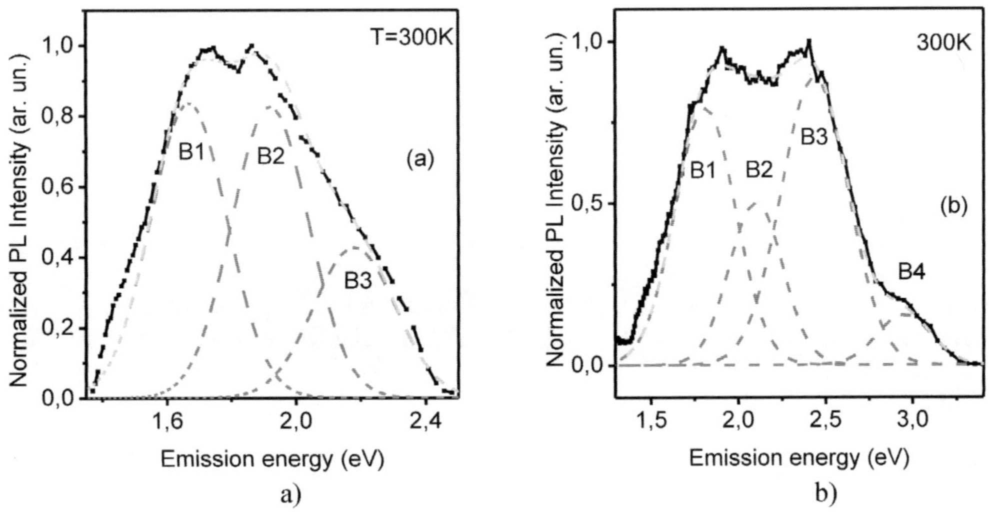

Figure 4 Deconvolution of PL spectrum excited by energy 2.54eV(a) and 3.82eV(b) measured at 300 K for the film grown with R=0.59.

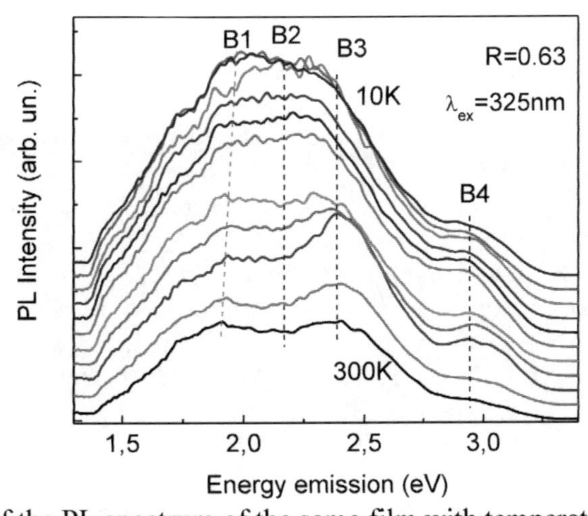

Figure 5 The evolution of the PL spectrum of the same film with temperature of measurement in the range 20-300K. Excitation energy is 3.82 eV.

Conclusions

Silicon-rich-silicon nitride films with different stoichiometry were fabricated by the PECVD method on the Si substrates The investigation of PL spectra using different excitation energies and temperatures gives the possibility to extract the information about the contribution of different radiative channels in the PL process. The optimal fabrication conditions for the bright emission were determined.

Acknowledgements The authors would like to thank the SIP-IPN, Mexico (projects 20160285), as well as the National Academy of Sciences of Ukraine and Center of National Scientific Research (CNRS) of France for the financial support.

References

[1] Belyi V I, Vasilyeva L L, Gennann R, Ginovker A S, Gritsenko V A, Repinsky S M, Sinitsa S P, Smirnova T P and Edelman F L, 1988 *Silicon Nitride in Electronics, Materials Science Monographs* vol.**34** (Elsevier Science Ltd, Amsterdam)

[2] Pavesi L and Lockwood D J 2004 *Silicon Photonics* Springer Berlin

[3] Sung G Y, Park N M, Shin J H, Kim K H, Kim T Y, Cho K S and Huh C 2006 *IEEE J. Sel. Top. Quant. Elect.* **12** 1545

[4] Wolkin M V, Jorne J, Fauchet P M, Allan G and Delerue C 1999 *Phys. Rev. Lett.* **82** 197

[5] Park N M, Choi C J, Seong T Y and Park S J 2001 *Phys. Rev. Lett.* **86** 1355

[6] Dal Negro L, Yi J H, Nguyen V, Yi Y, Michel J and Kimerling L C 2005 *Appl. Phys. Lett.* **86** 261905

[7] Comedi D Y, Zalloum O H, Wojcik J and Mascher P P 2006 *IEEE J. Sel. Top. Quant. Elect.* **12** 1561

[8] Pratibha Nalini R, Khomenkova L, Debieu O Cardin J, Dufour C, Carrada M and Gourbilleau F 2012 *Nanoscale Res. Lett.* **7** 124

[9] Delachat F, Carrada M, Ferblantier G, Grob J-J and Slaoui A 2009 *Nanotechnology* **20** 415608

[10] Keita A-S, En Naciri A, Delachat F, Carrada M, Ferblantier G and Slaoui A 2010 *J. Appl. Phys.* **107** 093516

[11] Delachat F 2010 Elaboration and Characterization of Si-licon Nanoparticles in Silicon Nitride for Photovoltaic Application (Ph.D. thesis) InESS-University of Strasbourg Strasbourg France

[12] Torchynska T V, Goldstein Y, Many A, Jedrzejewskii J and Kolobov A V 2003 *Microelectron. Eng.* **66** 83

[13] Khomenkova L Gourbilleau F Cardin J Jambois O Garrido B and Rizk R 2009 *J. Lumin.* **129** 1519

[14] Torchynska T V Casas Espinola J L Khomenkova L., Vergara E, Andraca Adame J A and Slaoui A 2015 *Mater. Scien. in Semicon. Proces.* **37**, 46-50 (2015)

[15] Park N M, Kim T S and Park S J 2001 *Appl. Phys. Lett.* 78 2575

[16] Rui Y, Chen D, Xu J, Zhang Y, Yang L, Mei J, Ma Z, Cen Z, Li W, Xu L, Huang X and Chen K 2005 *J. Appl. Phys.* **98** 033532

[17] Torchynska T V, Casas Espinola J L, Vergara E, Khomenkova L and Slaoui A 2015 *Thin Solid Films* **581** 65-69

IOP Conf. Series: Materials Science and Engineering **169** (2017) 012022 doi:10.1088/1757-899X/169/1/012022

Dependence of stoichiometry of lithium niobate nanocrystals on initial lithium to niobium ratios in the synthesis step

K Veenhuizen[1], G A Stone[2], B Knabe[3,4], K Buse[3,4] and V Dierolf[1]

[1] Department of Physics, Lehigh University, 16 Memorial Drive East, Bethlehem, PA 18015, USA
[2] Department of Materials Science and Engineering, Pennsylvania State University, Millennium Science Complex, University Park, PA 16802, USA
[3] Department of Microsystems Engineering (IMTEK), University of Freiburg, Georges-Köhler-Allee 102, 79110 Freiburg, Germany
[4] Fraunhofer Institute for Physical Measurement Techniques IPM, Heidenhofstrasse 8, 79110 Freiburg, Germany

Email: kjv212@lehigh.edu

Abstract. Ferroelectric nanocrystals show promise for application in forming hybridized nonlinear materials with liquid crystals. It is well known that bulk single crystals of lithium niobate ($LiNbO_3$) are most easily grown in a congruent (lithium-deficient) form but can also be grown in a stoichiometric form. This is controlled by the specific growth conditions and the stoichiometric ratio $\rho = M_{Li}/(M_{Li} + M_{Nb})$, where M is the molar fraction. This work explores the dependence of the stoichiometry of $LiNbO_3$ nanocrystals on the value of ρ in the synthesis step. Batches of $LiNbO_3$ nanocrystals were synthesized using a sol-gel method. The nanocrystals were analysed via SEM and Raman spectroscopy to gain information about their morphology, stoichiometry, defect content, and phase. For bulk crystals, previous work has demonstrated that the spectral widths of specific Raman modes strongly depend on ρ. For the nanocrystals, the Raman spectra indeed reveal that the resultant nanocrystal stoichiometry depends on the initial ρ used in the synthesis step. In addition, a close examination of the Raman spectra reveals the presence of an extra phase in batches with $\rho \geq 55\%$. Somewhat counterintuitively, this phase is identified by its Raman spectra to be $LiNb_3O_8$, a relatively lithium-poor phase compared to $LiNbO_3$. Avoiding this extra phase, we find that high quality, roughly spherical $LiNbO_3$ nanocrystals can be synthesized for ρ between 52 and 54%.

1. Introduction

Ferroelectric nanocrystals show promise in supplementing materials such as liquid crystals and photoconductive polymers [1]. A hybrid material is created where the nanocrystals improve an essential property of the primary material, for example the photoconductivity of a polymer [1]. An important aim is to improve some property of the primary material without any noticeable drawback due to the nanocrystals' presence [1]. For the nanocrystals to fulfill this purpose, there should be a uniformity of the resultant nanocrystal product in terms of morphology, stoichiometry, and phase.

In this work, a variety of $LiNbO_3$ nanocrystal batches made from different initial stoichiometric ratios were analyzed using SEM imaging and Raman spectroscopy. A fundamental concern of this work was to observe whether varying the initial stoichiometric ratio could control the final nanocrystal product. A question we seek to answer is, if the variation of the initial stoichiometric ratio influences the morphology, stoichiometry, or phase of the final product, then what initial stoichiometric ratio is ideal for applications?

2. Experimental Procedure

Nanocrystals of $LiNbO_3$ were synthesized using a double alkoxide sol-gel method that has been discussed in detail in previous publications [2]. Niobium ethoxide and lithium ethoxide are dissolved in ethanol and refluxed for 24 h. The solution is then slightly cooled and hydrolyzed by distilled water for an additional 24 h. The precipitate is dried and calcined. The product then underwent planetary ball milling in oleic acid before being suspended in water.

The initial stoichiometric ratio

$$\rho = \frac{M_{\mathrm{Li}}}{M_{\mathrm{Li}}+M_{\mathrm{Nb}}} \#(1)$$

is varied, and batches are created with values of ρ spanning 40 to 60% (40, 45, 50, 52, 54, 55, 56, 58, and 60%). Here M_{Li} and M_{Nb} are the molar fractions of the compounds.

The nanocrystals are characterized by SEM imaging and Raman spectroscopy. The morphology of the nanocrystals were observed via SEM. For Raman scattering measurements, nanocrystals are prepared in liquid suspension contained between glass slides as well as dried onto the surface of a glass slide. The confocal Raman microscope (figure 1) utilizes an Ar ion laser (Coherent Innova 70) operating at 488 nm, and light is collected in a backscattering geometry. Single mode fibers act as the excitation and collection pinholes. Spectra from dried nanocrystals are collected at a temperature of 4 K using an objective with numerical aperture 0.50. Spectra from suspended nanocrystals are collected using an oil immersion objective. The numerical aperture (NA = 1.32) of the oil immersion objective is sufficiently high to enable optical trapping of the nanoparticles. In such an optical tweezer, a dense collection of nanocrystals coalesces around the laser focus, resulting in a strong Raman signal from the cluster of nanocrystals.

The data is analyzed using the multipeak fit utility in IGOR Pro. Peaks were fit with Lorentzian profiles and a cubic polynomial baseline. The multipeak fit is restricted to peaks with Raman shift between 100 and 500 cm^{-1} to maintain high accuracy. The directional dispersion that is characteristic for the polar $LiNbO_3$ material [3] contributes to the asymmetric appearance of some peaks and somewhat reduces the accuracy of the fit. A lithium poor phase ($LiNb_3O_8$) was present in a few batches. It possesses a unique Raman spectrum, and so fitting was performed only when Raman spectra containing the $LiNbO_3$ signature were obtained, as the intention was to extract information from the best fit parameters on stoichiometry and presence of defects in $LiNbO_3$ nanocrystals.

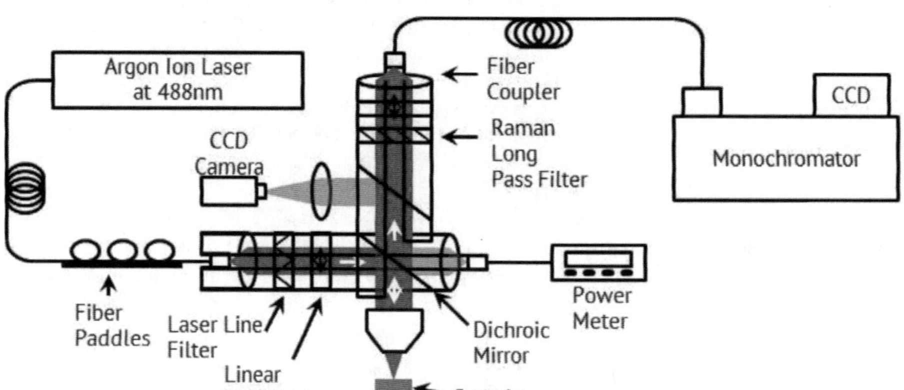

Figure 1. Schematic of confocal Raman microscope.

IOP Conf. Series: Materials Science and Engineering **169** (2017) 012022 doi:10.1088/1757-899X/169/1/012022

3. Results

3.1. Raman spectroscopy

In previous studies on bulk single crystal $LiNbO_3$, the full-width-at-half-the-maximum (FWHM) of the $E(TO)_1$ mode as well as the $A(LO)_4$ mode in regards to the lithium concentration were calibrated [4]. In this work, the nanocrystals in liquid suspension are oriented in various directions and interface related effects are more relevant [5], and hence we need to focus on the $E(TO)_2$ mode for which the directional dispersion contributes very little to its broadening [3]. Figure 2(a) shows the Raman spectra of nanocrystals in liquid suspension for a variety of the initial stoichiometric ratio while figure 2(b) displays the FWHM of the $E(TO)_2$ mode obtained from these spectra. Raman spectra were also collected for bulk single crystal z-cut congruent lithium niobate (CLN) and z-cut near-stoichiometric lithium niobate (nSLN). The FWHM of the $E(TO)_2$ mode for each of these bulk single crystals is drawn as a horizontal line for reference on figure 2(b). The 55% batch is seen to have the narrowest FWHM comparable to that of nSLN. All batches between 50 and 60% possess FWHM which fall between CLN and nSLN. The 40 and 45% batches show significant broadening as displayed in figure 2(a).

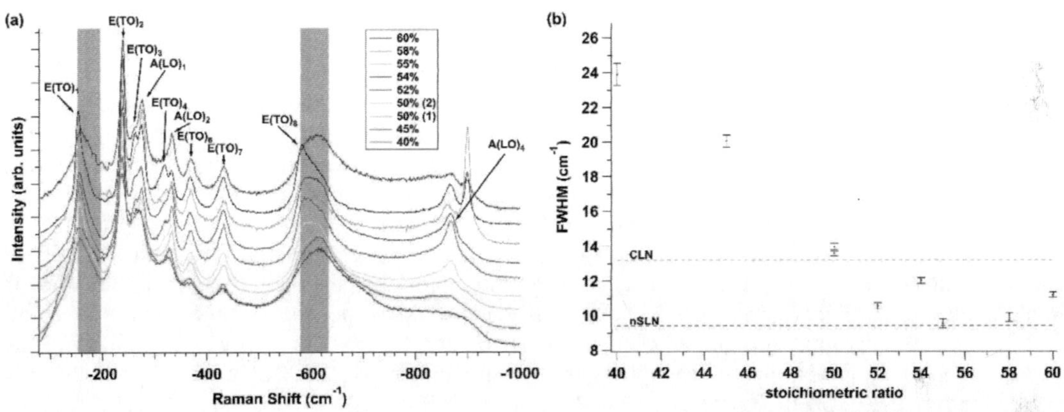

Figure 2. (a) Raman spectrum of liquid suspended nanocrystals for each initial stoichiometric ratio. The gray highlighted regions correspond to the wide $E(TO)_1$-$E(LO)_2$ and $E(TO)_8$-$A(TO)_4$ directional dispersion regions. (b) FWHM of $E(TO)_2$ mode vs initial stoichiometric ratio ρ with z-cut CLN and z-cut nSLN shown for reference.

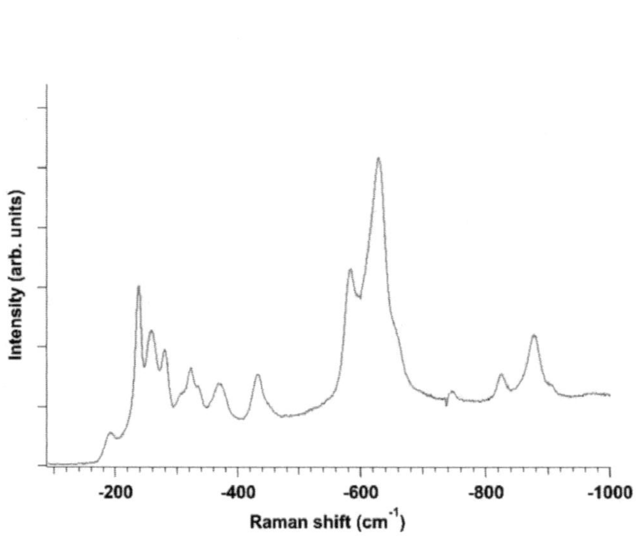

Figure 3. Raman spectrum of dried sample of nanocrystals taken from 58% batch at 4 K.

Inspecting the nanocrystal Raman spectra beyond the $E(TO)_2$ mode, there is a significant contribution from the E(TO) modes. Mode assignments are given in table 1 using the most recent literature data [6]. Directional dispersion is evident in the $E(TO)_1$-$E(LO)_2$ and $E(TO)_8$-$A(TO)_4$ regions leading to broadened asymmetric line shapes [3]. Some A(LO) modes are reduced like the $A(LO)_4$ mode, and A(TO) modes are largely suppressed, as evidenced by the absence of the $A(TO)_1$ mode.

There are differences between the Raman spectra for each initial stoichiometric ratio. For the 40 and 45% batches, the $A(LO)_4$ mode does not appear as a distinct peak. Looking at the $E(TO)_8$-$A(TO)_4$ directional dispersion region, there is a clear variation in the dispersion profile as a function of stoichiometric ratio. For values of $\rho \geq 55\%$, an intense extra mode appears at a higher energy than all the other modes at around 900 cm^{-1}. Other peaks not related to $LiNbO_3$ also appear, indicating the appearance of an additional phase. This phase is so dominant for the 56% batch that no data point is included in figure 2(b) for this batch.

A Raman spectrum collected from a dried sample of nanocrystals from the 58% batch is shown in figure 3(a). In such a dried sample, the orientation of the nanocrystals is random and cannot be controlled as in the case of liquid suspended nanocrystals trapped near the focus of the optical tweezer.

Table 1. Multipeak fit results for 55% batch and z-cut nSLN.

$\rho = 55\%$		z-cut nSLN	
Raman shift (cm^{-1})	Mode	Raman shift (cm^{-1})	Mode
156.4 ± 0.2	$E(TO)_1$	153.4 ± 0.1	$E(TO)_1$
239.1 ± 0.1	$E(TO)_2$	238.6 ± 0.1	$E(TO)_2$
264.1 ± 1.0	$E(TO)_3$	264.4 ± 0.2	$E(TO)_3$
276.8 ± 0.2	$A(LO)_1$	275.9 ± 0.2	$A(LO)_1$
320.3 ± 0.8	$E(TO)_4$	321.2 ± 0.2	$E(TO)_4$
333.3 ± 0.4	$A(LO)_2$	334.7 ± 0.1	$A(LO)_2$
370.3 ± 0.4	$E(TO)_6$	367.6 ± 0.2	$E(TO)_6$
		421.1 ± 0.2	$A(LO)_3$
432.6 ± 0.5	$E(TO)_7$	432.4 ± 0.2	$E(TO)_7$

3.2. SEM imaging

The SEM images in figure 4 reveal that the crystallization process improved with increasing values of ρ. The 55, 58, and 60% batches produced a uniform collection of near spherical nanocrystals. The 40,

45, and 50% batches all produced inhomogeneous collections of nanocrystals with jagged edges. The 40% batch produced the worst nanocrystal product, characterized by large clumps and lack of uniformity in the shape of the nanocrystals.

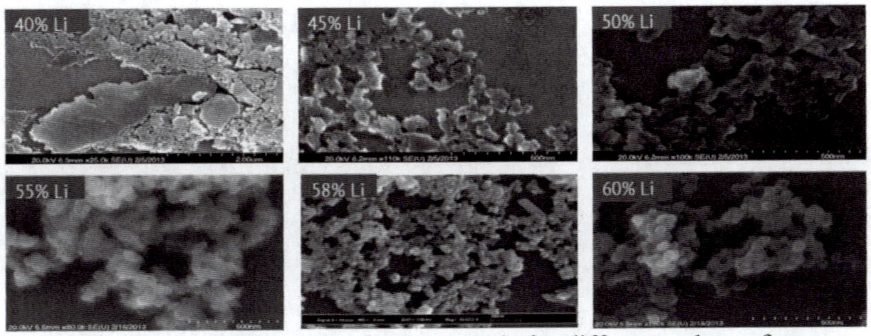

Figure 4. SEM images of nanocrystals for different values of ρ.

4. Discussion

The trend of figure 2(b) shows the FWHM hits a minimum around ρ_{min} = 55%. The batches with $\rho <$ 50% have the broadest peaks, indicating high defect content and off-stoichiometric nanocrystals. The FWHM of the 50% batch is near to that of CLN and 55% is near to that of nSLN, indicating the final stoichiometric ratio of the nanocrystals is less than the initial stoichiometric ratio before synthesis.

The Raman spectrum in figure 3 is a fingerprint of a collection of randomly oriented crystallites. There are a few clear distinctions between this spectrum when compared to the Raman spectra in figure 2(a) of nanocrystals in liquid suspension trapped by the optical tweezer. The most notable differences are in the 300-350 cm^{-1} and 580-630 cm^{-1} regions.

In the case of the optically trapped nanocrystals in liquid suspension, the directional dispersion in the $E(TO)_1$-$E(LO)_2$ and $E(TO)_8$-$A(TO)_4$ regions covers the entire dispersion range, implying the nanocrystals are aligning themselves at a variety of orientations in the optical trap. However, the difference when compared to the Raman spectrum of dried nanocrystals as well as the mode assignment of table 1 reveals a preferential orientation of the nanocrystals in the optical trap.

The absence of the A(TO) modes combined with the strong presence of the E(TO) modes suggests the preferred alignment is with the ferroelectric axis of the nanocrystals roughly parallel to the light propagation axis. The dominant contributor to the alignment of the nanocrystals is the induced dipole perpendicular to the spontaneous polarization of the nanocrystals, consistent with what has been observed in a previous study of $LiNbO_3$ nanocrystals in liquid suspension under application of an external electric field [7]. The observed alignment effect is most pronounced for those batches for which the nanoparticles were almost round and a competing alignment effect based on the geometric shape is absent.

The extra mode around 900 cm^{-1} apparent in the Raman spectra for $\rho \geq 55\%$ can be explained as a different crystal phase in the batch besides $LiNbO_3$. Figure 5(a) displays spectra where the extra modes, including the 900 cm^{-1} mode, are indicated. Comparing these characteristic Raman spectra with earlier work on thin films in Figure 5(b) [8] leads to the conclusion that this extra phase maybe $LiNb_3O_8$ [8]. This is somewhat counterintuitive because $LiNb_3O_8$ is lithium poor. It would be expected that when higher relative ratios of lithium to niobium are used in the synthesis step, a lithium rich phase like Li_3NbO_4 would appear. Further work (beyond the scope of this paper) is needed to study this point.

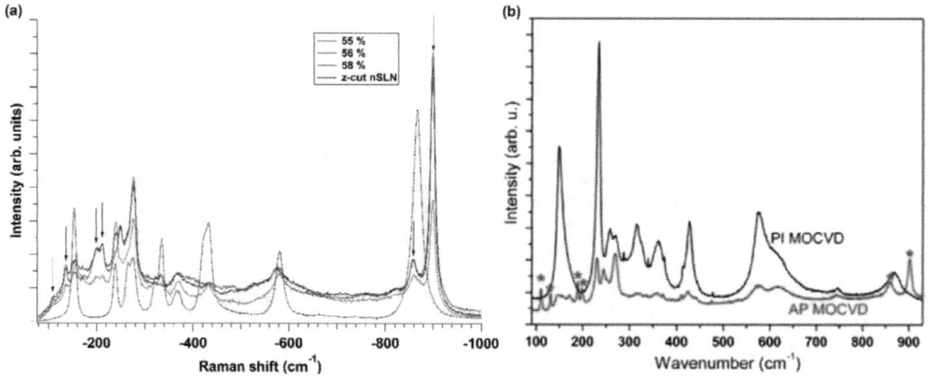

Figure 5. (a) Raman spectra revealing a combination of $LiNb_3O_8$ and $LiNbO_3$ with arrows pointing to Raman modes which can unambiguously be assigned to $LiNb_3O_8$ phase. (b) Figure 5(b) reproduced from Reference 8 showing Raman spectrum of thin film containing $LiNb_3O_8$ phase (in red).

5. Conclusion

It was observed that changing the initial stoichiometric ratio indeed varies the resultant nanocrystal product. Initial stoichiometric ratios with $\rho \geq 55\%$ produced uniform batches of spherical nanocrystals while batches with $\rho \leq 50\%$ produced non-uniform collections of nanocrystals. Furthermore, the FWHM of the $E(TO)_2$ was at a minimum for those batches with $\rho \geq 55\%$, indicating a near-stoichiometric $LiNbO_3$ nanocrystal product. Batches with the higher initial concentration of lithium to niobium ($\rho \geq 55\%$) also possessed an extra phase besides $LiNbO_3$. The 52 and 54% batches balance the avoidance of the extra phase while producing high quality, near-stoichiometric nanocrystals and are thus most suitable for application purposes.

Acknowledgements

We are grateful for the financial support of the NSF through the Materials World Network Grants (DMR-0602986 and DMR-1008075).

References

[1] Kurochkin O, Buchnev O, Iljin A, Park S K, Kwon S B, Grabar O and Reznikov Y 2009 *J. Opt. A: Pure Appl. Opt.* **11** 024003
Mansurova S, Meerholz K, Sliwinska E, Hartwig U and Buse K 2009 *Phys. Rev.* B **79** 174208
[2] Knabe B, Schütze D, Jungk T, Svete M, Assenmacher W, Mader W and Buse K 2011 *Phys. Status Solidi* A **208** 857-62
Knabe B, Assenmacher W, Mader W and Buse K 2012 *Phys. Rev.* B **86** 195428
Hirano S I and Kato K 1987 *Adv. Ceram. Mater.* **2** 142-45
[3] Yang X, Lan G, Li B and Wang H 1987 *Phys. Status Solidi* B **141** 287-300
[4] Schlarb U, Klauer S, Wesselmann M, Betzler K and Wöhlecke M 1993 *Appl. Phys.* A **56** 311
Ridah A, Bourson P, Fontana M D and Malovichko G 1997 *J. Phys.: Condens. Matter* **9** 9687
[5] Stone G and Dierolf V 2012 *Optics Letters* **37** 1032
[6] Margueron S, Bartasyte A, Glazer A M, Simon E, Hlinka J, Gregora I and Gleize J 2012 *J. Appl. Phys.* **111** 104105
Sanna S, Neufeld S, Rüsing M, Berth G, Zrenner A and Schmidt W G 2015 *Phys. Rev.* B **91** 224302
[7] Schütze D, Knabe B, Ackermann M and Buse K 2010 *Appl. Phys. Lett.* **97** 242908
[8] Bartasyte A, Plausinaitiene V, Abrutis A, Stanionyte S, Margueron S, Boulet P, Kobata T, Uesu Y and Gleize J 2013 *J. Phys.: Condens. Matter* **25** 205901

Institute of Physics
Dirac House, Temple Back
Bristol BS1 6BE UK

ISSN: 1757-8981
ISBN 978-1-5108-3673-0

CURRAN ASSOCIATES INC.
proceedings.com

41st Vietnam National Conference on Theoretical Physics (NCTP-41)

Journal of Physics: Conference Series Volume 865

Nha Trang, Vietnam
1 - 4 August 2016

IOP
Institute of Physics